D1098505

Colour: Art & Science

The Darwin College Lectures

Colour: Art & Science

Edited by Trevor Lamb and Janine Bourriau

CAMBRIDGE
UNIVERSITY PRESS

PUBLISHED BY THE PRESS SYNDICATE OF THE UNIVERSITY OF CAMBRIDGE
The Pitt Building, Trumpington Street, Cambridge CB2 1RP, United Kingdom

CAMBRIDGE UNIVERSITY PRESS
The Edinburgh Building, Cambridge CB2 2RU, United Kingdom
40 West 20th Street, New York, NY 10011–4211, USA
10 Stamford Road, Oakleigh, Melbourne 3166, Australia

First published 1995

Reprinted 1997

Printed in the United Kingdom at the University Press, Cambridge

A catalogue record for this book is available from the British Library

ISBN 0 521 49645 4 hardback
ISBN 0 521 49963 1 paperback

Contents

Introduction

Trevor Lamb and
Janine Bourriau

Although the idea of 'colour' may seem a simple concept, it conjures up very different ideas for each of us. To the physicist, colour is determined by the wavelength of light. To the physiologist and psychologist, our perception of colour involves neural responses in the eye and the brain, and is subject to the limitations of our nervous system. To the naturalist, colour is not only a thing of beauty but also a determinant of survival in nature. To the social historian and the linguist, our understanding and interpretation of colour are inextricably linked to our own culture. To the art historian, the development of colour in painting can be traced both in artistic and technological terms. And for the painter, colour provides a means of expressing feelings and the intangible, making possible the creation of a work of art.

To comprehend the many aspects of colour, we must travel all the way from the hard sciences to the fine arts. Although these so-called 'two cultures' have frequently been seen as antithetical to each other, an exciting view to emerge from this volume is the convergence of ideas from disparate approaches. In the field of colour, the arts and the sciences now travel in unison, and together they provide a rich and comprehensive understanding of the subject.

On the scientific side, our path leads from the physics of light, through the biology of the nervous system and the intricacies of the mind, and on to the living organism in its environment. In the humanities, we traverse the history of colour in art, assess the role of our own culture and language in colour concepts, and experience the guiding principles of colour for a contemporary painter. Our path in these travels comprises essays by eight of the most eminent practitioners in their respective fields.

David Bomford, the Chief Restorer of Paintings at the National Gallery in London, introduces us to a history of colour in Western art. In the fourteenth and fifteenth centuries, two colour methodologies were introduced into painting. In 1390, Cennino described a system in which colours were used in their purest (saturated) form in the deepest shadows, and were then lightened progressively (de-saturated) with white in the lit areas, finishing with highlights of pure white. Half a century later, Alberti advocated a system in which the pure colour could either be modelled up with white, or instead down with black, in the lighter and darker regions respectively, with the pure colour occupying a position somewhere in the middle. Although Cennino's system remained the standard that characterizes tempera paintings, Alberti's more subdued method can be seen in the work of some painters of the time. A further theoretical insight attributable to Alberti was an elaboration of the concepts that we now call colour contrast and complementary colours, ideas that have been employed by successive generations of painters.

It was not only progress in ideas and theories that drove the evolution of painting, but technology also played a powerful part. The spread of oil-based painting in the fifteenth century provided artists with a whole new repertoire: a glossy, viscous medium, offering a wide range of colours and tones. Much later, in the eighteenth and nineteenth centuries, the blossoming science of chemistry provided a range of new synthetic pigments, which were rapidly adopted by the painters of the day. Not only has modern scientific analysis of paintings and pigments given us a fuller understanding of the techniques and materials used by painters, but it has also provided a formidable weapon in determining authenticity: a number of paintings have been shown to contain pigments not invented until after the death of the supposed painter.

How do the ideas and discoveries of the great painters of the past affect and influence a painter of the present? In this we are fortunate to be guided by one of Britain's most acclaimed contemporary artists, *Bridget Riley*. A painter has two quite different systems of colour to deal with: firstly, perceptual colour, i.e. colour as it is perceived by the observer viewing a natural scene; and secondly, pictorial colour, i.e. the use of the pigments from the palette to fabricate the appropriate sensation to the viewer. Over the centuries, the investigation of pictorial colour, by a relatively small number of painters, has led to some remarkable artistic adventures.

Titian, arguably the greatest colourist, introduced an incredible innovation by rendering colour inseparable from the forms depicted; he developed a style in which colour was used as a single element to shape and form the painting. A continual source of fascination is the way in which Titian allows a colour sensation to run through an entire composition, turning here warmer and there more subdued, subtly changing in hue, emerging now and then in strong tones, but winding through the work as a continuum. This development was to put European painting on a unique road – one reaching to the Impressionists and Post-Impressionists. The importance of Titian's discovery, and its subsequent evolution through the hands of later painters, is traced out by Bridget Riley with the passion of a practitioner of the art.

But what are the physical mechanisms that underlie the very existence of light and colour? *Malcolm Longair*, Jacksonian Professor of Natural Philosophy in the University of Cambridge, unravels our present understanding of light and colour as an historical development covering the last 400 years. The law of the refraction of light was discovered by Snell and Descartes early in the seventeenth century, and provided an explanation for the origin of rainbows. But the addition of colour to the rainbow did not come until later that century, when Newton's experiments with a glass prism showed that white light was composed of all the colours of the spectrum. In a remarkable burst of genius, the 22 year old developed the calculus in mathematics, the theory of colour in optics, and the theory of gravity in astronomy.

It was not until the nineteenth century that Thomas Young and Hermann von Helmholtz discovered the trichromacy of colour vision: the fact that any perceivable colour could be created by mixing together lights of only three different colours – this was a discovery not so much about light itself, but about the mechanism of human colour vision. Their ideas were put on a quantitative basis by James Clerk Maxwell, whose experiments on the mixing of colours form the basis of our current understanding.

How does the light that reaches our eye trigger a neural response, and what is the biological mechanism underlying the trichromacy of colour vision? The remarkable discoveries of recent years are described by *Denis Baylor*, Professor of Neurobiology at Stanford University. The retina lining the back of the eye contains sensory receptor cells: the rod and cone photoreceptors. The cone cells, which mediate daylight vision, comprise three types, sensitive to light in the red, green and blue portions of the spectrum. Electrical recordings from these microscopic cells, only one-thirtieth of the diameter of a human hair, have now revealed the complex chain of biochemical reactions that endow our photoreceptors with their exquisite sensitivity to light. The spectral properties of individual cones have been examined, and they are found to account precisely for the measured colour performance of the human observer – confirming exactly the ideas and observations of Young, Helmholtz and Maxwell last century. The human retina, although very thin, comprises a complex piece of computational machinery. Recent research has discovered the sequence of information-processing steps that the retina performs on colour and luminance signals. Thus, we now understand in considerable detail both the initial conversion of coloured light into a neural signal, and the subsequent processing of this information within the human eye.

But how are the neural signals that are sent from the eye processed and interpreted by the human brain? *John Mollon*, Reader in Psychology in the University of Cambridge, analyses the two separate colour systems that we are now known to possess – an ancient system shared by all mammals, together with a system that evolved more recently in our primate forebears. The older system provides dichromatic colour vision, by comparing responses from two classes of cones, while the newer system adds a further dimension to our colour vision, by bringing in the signals from the most recently evolved set of cones, sensitive at the red (or long-wavelength) end of the spectrum. By studying the performance of normal observers, as well as certain colour-anomalous subjects (including red/green deficient males and their mothers), it has been possible to determine the role and the relative importance of the two systems in different colour-recognition tasks. The ancient and modern systems, perhaps surprisingly, do not appear to correspond to the colour-opponent red/green and blue/yellow channels proposed by Ewald Hering.

In Nature, colour may be a source of great beauty to the beholder, but to a plant or an animal it is most often a means of survival. *Peter Parks*, a founder of Oxford Scientific Films and Image Quest, illustrates the magnificence of colour in Nature with stunning examples. Natural colour arises from a diversity of mechanisms, often associated with distinct functions. Dyes and stains are used by many creatures, frequently for camouflage. Structural colour – typified by the metallic sheen of a Mallard duck's blue speculum – is generally used for the intense colours that announce the presence of an individual, as for example in a mating display. Colour can also be used for warning, as in the poisonous fire-bellied toad, or even for mimicry. Less commonly, animals can dispense with colour altogether and become transparent, or, as in the squid, they can actively control skin colour for the purpose of camouflage, or for attraction and mating. Not only are the mechanisms and purposes multitudinous, but the effects are spectacular.

The discourse progresses next to the experience of colour by the observer, and to the meanings and ideas associated with particular colours. Why, for example, do we (in a Western culture) think of blue as cool, and red as warm? *John Gage*, Head of the Department of History of Art in the University of Cambridge, addresses these issues in terms of the relation between colour and culture. Even our concept of the number of different colours making up the rainbow seems to have arisen from a cultural factor – from Newton's desire to sustain the classical analogy between musical and optical harmony, between his scale of colours and the musical octave.

In some cultures, when specifying a colour, the texture of the object is more important, and the hue less important, than is the case in our own culture. And within Western culture, the enormous increase in interest in highly contrasting hues that was evident in the arts around the turn of this century probably arose from a combination of psychological and technological developments in society. So, colour and culture are inextricably linked.

All our concepts of colour are expressed in language. *John Lyons*, Master of Trinity Hall, Cambridge, explores the limitations placed on our understanding of colour, and on the naming of colours, by our use of language. No easy equivalents can be made between the colour-terms of different languages, and in English, a language rich in colour-terms, many are specific to particular contexts. A distinction needs to be made between the use of colour-terms to qualify an object (descriptive use) and the use of colour-terms to refer explicitly to a colour (referential use). In some languages, referential use is much less common than in languages such as English, and indeed many languages have no word for colour itself.

This leads him on to examine the influential thesis of Berlin and Kay that all languages recognise the same 'basic' colour-terms. He uses the examples of the modern Hanunoo language of the Philippines and Ancient Greek to show how words *translated* as simple colour-terms may include, in their original context, properties such as texture and luminosity. Thus *chloros* in Greek is usually translated to mean *green* in English, yet its meaning appears to refer to the freshness or moistness of green foliage, rather than to its colour. A difference in colour perception is not the cause of this language difference; the Greeks were not colour blind. It is instead a matter of language and culture whether the hue of an object is regarded as important, and indeed whether there is a name for it at all.

Thus the perception of *colour* is in the mind of the beholder, yet *colours* as we know them are a product of our language and our culture.

1   The History of Colour in Art

David Bomford

Introduction

It is almost impossible for us to know just how colourful the art of the past was. We can hardly begin to imagine how extraordinarily sumptuous mediaeval and Renaissance churches and palaces appeared – with their wall paintings, tapestries, painted architectural ornament, precious metals, enamels, and every kind of brilliant artifact. Any history of colour in art can only be partial, because so much art and so much colour in art has either perished or has survived only in a much changed form.

We have only to think of sculpture, for example, to see how our imagination fails us. With classical Greek and Hellenistic marble statues, it always comes as something of a shock to realise that they normally had realistically coloured lips, eyes, hair and clothes. We now usually imagine the antique through the practices of Renaissance and neo-classical artists, who saw Greek sculptures already stripped by time or the hand of Man of all their painted decoration. Polychromy of stone sculpture was certainly normal right up to and through the Romanesque and Gothic. It was only in the Renaissance – by a combination of mistaken interpretation of the nature of classical sculpture and a genuine interest in the natural textures and colours of materials that polychromy of stone sculpture died out. Very little intact medieval polychromy of carved stone survives, but where it is found – as in the great west portal of the collegiate church in Toro in north central Spain (The Portada de la Majestad, dating from the late thirteenth century), hidden for centuries under many layers of later polychromy – then the impact is astonishing.

Polychromy of wooden sculpture, of course, continued right through the Baroque and beyond. But a moment can be identified when one artist decided to strike out in a different direction. In mediaeval and early Renaissance Germany, it was usual for limewood sculpture and altarpieces to be painted. Then, in 1490–2, Tilman Rimenschneider, the greatest of all the limewood sculptors, made the first known limewood retable in monochrome – the Munnerstadt altarpiece, which now survives only in fragmentary form. It was not entirely uncoloured: the pale wood was stained a little darker than its natural colour. But from then on, two traditions developed side by side and the more expert carvers revelled in the new unpainted freedom to show off the fineness of their carving and the subtle wood textures that they could produce.

There is an ironic sequel to Riemenschneider's making of the Munnerstadt altarpiece: eleven years later, the parish decided it was too plain, and in 1503 commissioned Veit Stoss to paint it in the old manner. His polychromy stayed on it for three centuries or more; then – we do not know precisely when – it was stripped off, a process that undoubtedly removed the original glaze beneath as well. Colour in art can be as untrustworthy as it is vulnerable.

Il libro dell' Arte: colour combinations based on pure pigments

Our best chance of constructing a coherent history of colour in art is to look at representational painting in all its forms from late mediaeval times to the twentieth century; the best place to start is with the most famous and influential treatise in the history of painting, *Il libro dell' Arte* written around 1390 by the Tuscan painter Cennino Cennini.

In his book, Cennino gives detailed instructions on the preparation of materials for painting in fresco and on panel. Learning to draw is important, he says, but working up the colours and painting with them is the 'glory of the profession'.

Cennino clearly describes systems of colour for depicting flesh, draperies, buildings and landscapes. Importantly, they are systems devised for painting in fresco and egg tempera – opaque, quick-drying media that were used in simple, direct techniques.

To paint faces, for example, Cennino describes how the flesh must first be underpainted with the pale green earth, terre verte. The pink flesh tones were then hatched or painted thinly on top, working in progressively paler shades from shadow to light; the green was allowed to show through in the half-tones and nicely imitated the pearly tones of real flesh. Today, many such faces are worn and damaged and the green has become too prominent. Cennino was very strict about this correct sequence for painting flesh: 'some begin by laying in the face with flesh colour – then they shape it up with a little verdaccio [*brown-green shadow colour*] and flesh colour, touching it in with some highlights and it is finished. This is a method for those who know little about the profession.' Later, in a famous passage, he recommends the pale yolk of a town hen's egg for painting the faces of young people with cool flesh colours, but the darker yolk of a country hen's egg for aged or swarthy persons.

Cennino's methods for painting coloured draperies were also highly specific, and formed the basis for painting the clothed figure right through the quattrocento and beyond. Essentially, colours were used in their pure form in the deepest shadows and then lightened progressively with white towards the lit areas, finishing with highlights of pure white. For its time, it was a remarkably successful scheme, but there were problems with it.

First, by placing the purest and most powerfully saturated colour in the deepest shadows and progressively desaturating it towards the lights, the shadows appeared to advance and the lights appeared to recede – the very opposite of the desired effect. Secondly, the relative brightness of the pure colours was very variable: this could lead to the unbalancing of compositions in which the brighter draperies, such as the yellows, stood out much more prominently than the darker ones, such as the blues. For this reason, painters often attempted a balance of symmetry, arranging their bright colours in pairs around a central axis – a scheme now termed *isochromatism.*

With the Cennino system, we thus have a series of colour combinations based on, and pre-determined by, the pure forms of the available pigments. Cennino describes the preparation of pigments from a variety of sources, both natural and artificial. Such colours could be readily available and inexpensive, or rare and cost a fortune: in the latter category, the best-known is ultramarine blue (literally, from 'over the sea', since it was then found only in Afghanistan), extracted from the semi-precious stone lapis lazuli and invariably (correctly) described as more expensive than gold. In late mediaeval times, when paintings were valued by the worth of their materials as much as the skill of their execution, the purest ultramarine was reserved for painting the Virgin's mantle and often costed separately in painter's contracts.

In such a painting as Lorenzo Monaco's *Coronation of the Virgin* (Figure 1), many of the available pigments are seen at full strength or mixed with white – coloured earths alongside ultramarine blues and the very beautiful lead-tin yellow. There was no green available powerful enough to compete with these strong colours and so areas of green tended to be mixtures of blue and yellow. Reds might be vermilion (an

Figure 1 Monaco, *Coronation of the Virgin.* (The National Gallery, London.)

artificial form of the mineral cinnabar) or lake pigments – made by adsorbing natural dyestuffs on to a white base.

Lake pigments were sometimes very fugitive when exposed to light and we have in this painting a dramatic example of how a colour can change and alter the appearance of an entire work. The Virgin's robe was originally a deep mauve-pink, not white as it appears today. The evidence is satisfyingly exact. A minute sample of paint from the main area of the robe, seen in cross-section, shows many colourless particles which were once mauve. A sample taken where the robe passes below some mordant-gilding – where it was protected from the light – shows the particles still with their original colour.

Painters sometimes made precise decisions about their pigments which suggest a sophisticated knowledge of material properties. In his Santa Croce altarpiece, the Sienese painter Ugolino di Nerio deliberately chose azurite blue for its greenish tonality, instead of the ultramarine that might have been expected for such a prestigious commission: even the Virgin's robe is azurite. This conscious choice set up a whole series of sub-tle colour contrasts and harmonies that mark out Ugolino as one of the most innovative colourists of the trecento.

Cennino's treatise mentions one other colour system for draperies that was used widely right through the Italian Renaissance and later became firmly associated with Michelangelo and the Mannerist painters. This was the system of *cangiantismo* – of showing shot or *cangiante* fabrics that appear to change between the lights and the shadows. Cennino lists a number of colour combinations appropriate for *cangiante* effects and one of them can be seen in in Nardo di Cione's *Altarpiece: Three Saints*, painted around 1365. One saint has a greenish robe, yellow in the high-lights and ultramarine blue in the shadows, painted just as Cennino described.

This is quite a subdued example of *cangiantismo* – much more brilliant hue-shifts can be seen in, for example, Mantegna's *Virgin and Child with Saints*, painted towards the end of the fifteenth century in tempera on canvas, in essentially Cenninian colouring.

Della Pittura: the use of black and white and colour chords

The colour systems described by Cennino set the basic pattern for tempera painting in the quattrocento. Then, in 1435–6 Leon Battista Alberti published his *Della Pittura*, most famous for its revolutionary systematic description of single-point perspective. Suddenly, three-dimensional space could be realistically drawn on a two-dimensional surface – and representational art was changed forever.

Alberti also had a number of things to say about colour. His actual definition of colour primaries – which he listed as blue, red, green and earth colour – was based on the old Aristotelian tradition and was not very helpful. His remarks on modelling and relief are much more useful. His method assumes that light will strike an object from one side only, forming clearly defined areas of light and shade: 'Note the middle of it with a very fine line so that the method of colouring it will be less in doubt.' A clear example of this subdivision into light and dark halves is Fra Angelico's predella panel of the *Miracle of St Nicholas*, painted in the late 1430s.

Then we come to the manipulation of colour itself: 'the same colour, according to the light and shade it receives will alter its appearance – we must consider how the painter ought to use black and white ... with great restraint you will commence to place the black where you need it and at the same time oppose it with white.'

Alberti's system is unmistakably different from Cennino's. Cennino constructs forms simply by modelling up from the pure colour with white: Alberti recommends modelling up with white and down with black by equal amounts, the pure colour now being positioned somewhere in the middle. Alberti was, however, very conscious of the fact that both white and black desaturate and diminish colours and exhorted his readers to be sparing with them: 'I cannot overemphasise the advantage of the frugality to painters. It would be useful if white and black were made from those large pearls that Cleopatra destroyed in vinegar so that painters would be miserly with them and their works would be truthful, sweet and pleasing.'

The influence of Alberti's system on tempera painters of that period is debatable. Certainly Cennino's method remained the standard one, and without detailed analysis it is difficult to be sure just which painters mixed black in with their shadows. However I would mention just one example in which both systems seem to be present on the same panel. In the *Trinity Altarpiece* (Figure 2), mainly by Pesellino but finished by Fra Filippo Lippi after Pesellino's death in 1457, the orange-red cloak of the right hand saint (ignoring the bottom half which is a later restoration) appears to be up-modelled with white by the Cennino method, while the red robe of the saint towards the left appears to be up- and down-modelled with white and black according to the Alberti method. Notice,

Figure 2 Pesellino and Fra Filippo Lippi, *Trinity Altarpiece.* (The National Gallery, London.)

how much more subdued is the Albertian colouring – reminding us of his strictures about frugality in the use of black and white.

Alberti explored another aspect of colour in *Della Pittura* and this was what we now call the colour chords, colours that enhance each other when placed together, a phenomenon that in later centuries developed into a theory of complementary colours: 'There is a certain friendship of colours so that one joined with another gives dignity and grace.' Amongst other colour combinations, he says, 'rose near green and sky blue gives both honour and life'. This is exactly the sort of colour combination that Alberti would have seen and admired in Masaccio's fresco paintings in the Brancacci Chapel, and exactly this combination of colours is found in Cosimo Tura's *Roverella Altarpiece* (of around 1475) with its fantastic architecture mirroring the pink and green robes of the angels and the whole scene improbably constructed against the deep blue sky. Tura was just the sort of academic painter who might have read Alberti.

The development of oil painting

Alberti's colour system did offer painters more control of their pigments in tempera, but the whole technical basis of painting was about to change. The development of oil painting in northern Europe introduced a use of colour that was eventually to sweep away egg tempera completely.

By the early fifteenth century, oil for painting had been around for two centuries or more. Indeed, there are grounds for thinking that oil was the indigenous painting medium in northern Europe, only temporarily displaced during the fourteenth century by the Italianate egg tempera technique of the International Gothic style. By the 1420s and 30s, Netherlandish painters such as Van Eyck and Robert Campin were using it in an extraordinarily refined way for panel painting. They had not abandoned the use of egg entirely – most of these panel paintings have underpaints of egg and finishing layers of oil – a fact that had been long suspected but only recently proven using microanalytical techniques.

The properties of oils, such as linseed, walnut or poppy, that make them so desirable for painting, are that they are glossy, viscous, slow-drying and highly refractive. Paint layers can be made thick or thin, opaque or transparent and individual brush strokes can be sharp and dramatic or blended imperceptibly until they disappear. Most importantly for colour, the same pigments that are opaque and high-key in egg can, in oil, become rich, semi-transparent glazes.

The implications for colouring within paintings were immediately apparent. Shadows no longer had to be unrealistic pure pigments or dully down-modelled with black. Now they could have infinite subtlety; they could be dark and full of colour at the same time. Half tones could be modelled with infinite softness. Highlights could be lustrous or dazzling.

The transition from egg tempera to oil as the principal panel-painting medium was a much faster process in northern Europe than in Italy. By the mid fifteenth century, the switch from egg to oil was almost complete in the Netherlands; in Italy it had only just begun. The precise mechanism by which the assimilation of oil painting into Italian practice occurred is still not clear. Vasari's account of Antonello da Messina travelling to Flanders to learn the technique from Van Eyck is chronologically impossible, but clearly Antonello was an early practitioner.

One recent clue has been provided by the analysis at the National Gallery of the Ferrarese painter Cosimo Tura's *Allegorical Figure* (Figure 3), painted for the studiolo at Belfiore in the late 1450s. The lowest layers are of egg tempera and may relate to a different composition; but the layers corresponding to the finished composition are solely of oil, used in an unequivocally Netherlandish technique of opaque underlayers modelled with glazes. The intricate brocade sleeve is strikingly similar to that of one of the kings in Rogier van der Weyden's *Columba Altarpiece* of around 1455 – and indeed, other paintings by Rogier were known to be in Ferrara by that date. Some historians have even cited the possibility that Rogier himself may have visited Ferrara and instructed Tura on a pilgrimage to Rome in 1450, but this cannot be verified.

The transition from egg to oil in Italy is difficult to chart, except by microchemical analysis of the paintings themselves, because many painters used oil in much the same way as they had used egg, and there is little visible difference in the appearance of their works.

But, essentially, by 1500 the predominant painting medium was oil, and its versatile properties prompted painters to explore a whole range of colouring systems that led European painting in various directions over the following centuries. Let us look briefly at one or two examples.

Figure 3 Tura, *Allegorical Figure.* (The National Gallery, London.)

Chiaroscuro and *sfumato*

The essential truth of Alberti's observations on the modelling of relief persisted and became the basic *chiaroscuro* mode of colouring in the sixteenth century. In Sebastiano del Piombo's *Raising of Lazarus* (1518), we can see how areas of high-key colour are set against sharp, deep shadows. The whole effect is highly contrasted, precise and crisp, but essentially fragmented.

A century later, Caravaggio was to refine the *chiaroscuro* mode into something altogether more atmospheric. Meanwhile, Leonardo da Vinci developed his own method of achieving a magical dark harmony at first glance derived, but in reality quite different, from *chiaroscuro*, this was

Figure 4 Leonardo da Vinci, *Adoration of the Kings*. (Galleria degli Uffizi, Florence.)

the colouring mode known as *sfumato* (literally, smoke-like) in which finely modulated forms emerge from soft shadows.

In such paintings as the *Virgin of the Rocks* (of about 1508) the subdued tonality is based on an extensive undermodelling in monochrome black, brown and grey. Such an undermodelling had been used by Piero della Francesca in his oil paintings from the 1470s, but Leonardo went much further. His colours were thinly applied and low in key and, muted by the deep monochrome below that, virtually eliminated any luminosity from the white ground. We can see this dark underpaint directly in the unfinished hand of the angel where it touches St John's back. We can see it completely uncovered in the unfinished *Adoration of the Kings* (Figure 4) of 1482.

Cangiantismo: the Sistine Chapel

Almost exactly contemporaneous with the *Virgin of the Rocks*, but as different as it is possible to be is Michaelangelo's Sistine Chapel vault, revealed in its astonishing colours by recent cleaning. The scheme of colouring here is *cangiantismo* on an unprecedented scale. There can be no doubt that these are the brilliant colours that Michelangelo intended – needed, indeed – to make the dark vault of the chapel light up and to render the individual scenes visible from far below. Carried out in true fresco, the colouring is essentially derived from Cennini – up-modelling, with *cangiante* draperies everywhere. The extraordinary turquoise-violet or orange-blue hue-shifts anticipate the complementary contrasts of the Impressionists by three and a half centuries.

Titian: *disegno* versus *colore*

Any history of colour in art must take as its central point the emergence and flowering of the sixteenth century Venetian school culminating in Titian, generally considered the greatest colourist of all. It is no surprise that Venice, the centre of the pigment trade in Europe, should foster a group of painters who delighted in colour and filled their paintings with an abundance of the finest pigments available. We might take as our first example *The Incredulity of St Thomas* by Cima (painted around 1500), which has been found to contain virtually every pigment available at that date – including some which were rarely used, such as orpiment, realgar and haematite. Moreover, he used different grades of azurite and ultramarine for different colour effects, and highly unusual organic glazing colours. The general sense is one of great sophistication, both in pigment

Figure 5 Titian, *Bacchus and Ariadne.* (The National Gallery, London.)

use and colour composition. There is only one repeated colour in the entire work – the green worn by St Thomas and the apostle at the far right.

Like Michelangelo's Sistine Chapel ceiling, Titian's *Bacchus and Ariadne* (Figure 5) has been revealed in its true colours by cleaning in recent years. It is one of the most radiant images in art, made so by the extraordinary range and purity of the pigments. Almost every pigment available in Venice is here, all of unparalleled quality and used at full strength. The ultramarine in the sky is the purest found in any painting yet examined at the National Gallery.

What is so striking about the use of strong competing colours here is that the composition could so easily have fallen apart visually, like the Sebastiano *Raising of Lazarus*. Yet Titian has achieved an overall harmony effortlessly, knitting everything together with a subtle framework of quiet greens and earth colours.

Titian's paintings of this period are complex in their layer structure, largely because he worked out many of the details of his compositions only at the painting stage. Let us look at the figure of Ariadne, for example, standing against the sea and sky, the vivid red scarf across her shoulder. You or I – or indeed Michelangelo – would probably have drawn it all out and then carefully filled in the colours to each area. But not Titian – he made it up as he went along: a cross-section shows that he painted the sea first, Ariadne's bare shoulder next and finally the scarf in the purest vermillion on top of that.

Titian's paintings became, more and more, acts of spontaneous creation on the canvas itself. It was this tendency that led to a famous instance of the perennial *disegno* versus *colore* quarrel. Ever since Aristotle, scholars had debated whether drawing or colour was more important in painting. Cennino and Alberti had sensibly allowed them equal importance. But Vasari could not resist joining in the fray – especially if, as a good Tuscan, he could score a few points off a Venetian.

In his Life of Titian he described a conversation with Michelangelo about Titian's method in some paintings they had just seen: 'Buonarroti commended it highly, saying that his colour and style pleased him very much, but it was a shame that in Venice they did not learn to draw well.' Vasari was especially scathing about Giorgione who, he said, 'failed to see that, if he wants to balance his compositions … he must first do various sketches on paper to see how everything goes together.'

The opposition of *disegno* and *colore* was not simply drawing versus colour: as we have seen, Michelangelo, the greatest exponent of *disegno*, was also capable of astonishing colour. It is, rather, the method of creation: Titian's habit of creating his compositions directly in paint on the canvas is the essence of *colore*.

We can only imagine how much more aghast Vasari would have been to read Palma Giovane's celebrated account of the older Titian at work on such paintings as the late *Death of Actaeon*.

> He used to sketch in his pictures with a great mass of colours as a bed or base for his compositions … then he used to turn his pictures to the wall and leave them there without looking at them, sometimes for several months. When he wanted to apply his brush again, he would examine them with the utmost rigour, as if they were his mortal enemies to see if he could find any faults. Then he gradually covered these forms and in the last stages he painted more with his fingers than his brushes.

Titian paved the way for the great *alla prima* painters of the Baroque: Rubens, Velázquez and Rembrandt. All three fell under his spell, Rubens making copies of the great Bacchanals in Madrid, Velázquez seeing his paintings daily at the Court of Philip IV and Rembrandt basing his 1640 self-portrait on Titian's *Man with a Blue Sleeve*, seen briefly in Amsterdam.

Each of them developed colour in highly personal ways. Rubens persisted with white grounds in such luminous and brilliant works as the 1609 *Samson and Delilah*, painted immediately after his return from Italy and influenced strongly by the jewel-like colours of Adam Elsheimer.

Rembrandt and Velázquez painted with much more limited palettes but with a sophisticated control over their materials that is only now being discovered. They used both light coloured and dark tinted grounds and experimented with unorthodox pigment mixtures to achieve extraordinarily subtle effects. In his wonderful portrait of his wife *Saskia in Arcadian Costume*, Rembrandt has added the highly unusual azurite to many parts in order to give a cool greenish tint to the whole picture. And Saskia's waistband is a notable piece of Rembrandt bravura, testing the properties of oil paint to their limit, challenging us to see the illusion beyond the reality.

Throughout the seventeenth century, great painters all over Europe developed their individual styles and produced masterpieces of astonishing diversity. The *colore* versus *disegno* debate raged on, with the battles between the Poussinistes and the Rubenistes in the French Academy in the 1660s and 1670s – battles that had more to do with spontaneity and formality than with colour and drawing. Technically, the processes of painting varied little from place to place or from painter to painter. It is worth remembering that, despite the extraordinary richness and variety of seventeenth-century paintings, all were produced using essentially similar materials and techniques.

The introduction of synthetic pigments

In terms of painting materials, the modern era began in 1704, with the invention of Prussian Blue. This is a date that every student of painting techniques knows as a *terminus post quem* and because it marks the beginning of the synthetic pigment industry that was to lead to a complete rethinking of the artist's palette in the early nineteenth century. Prussian blue became widely used within twenty or thirty years. Canaletto, who in his earliest works used ultramarine for his skies, was certainly using Prussian Blue by the time he painted the *Stonemason's Yard* in the late 1720s.

The precise dating of pigment inventions gives us a formidable weapon in the matter of determining authenticity. A number of apparently old paintings have been betrayed by the presence of Prussian Blue: and *Entrance to the Cannaregio* (Figure 6) once firmly attributed to Francesco Guardi was hurriedly relabelled 'Imitator of Guardi' when it was found to contain Cobalt Blue, invented nine years after Guardi's death.

In the first three decades of the nineteenth century an extraordinary number of new pigments appeared, the direct result of a rapidly expanding chemical industry. The discovery of cobalt, chromium and cadmium, the synthesis of artificial ultramarine in 1826 (following a competition sponsored by the French government) and the first synthesis of alizarin,

Figure 6 Imitator of Guardi, *Venice: Entrance to the Cannaregio.* (National Gallery, London.)

the red colouring of the madder plant in 1868, were just some of the notable landmarks in the history of modern pigments.

Nineteenth century painters adopted these new materials as soon as they became commercially available. Moreover, the development of the flat-ferrule brush, and – even more important – collapsible metal paint tubes made of lead or tin (available from about 1840), transformed painting practice. Brilliant, mostly stable colours were plentifully available and it was now easier than it had ever been for painters to work out of doors.

The Impressionists

All painters benefited from these developments, but it is the Impressionists that we think of automatically, and I want to look now at the use of colour by the Impressionist group. So much has been written, so many myths have been constructed about their art that it is difficult to know where to begin. Perhaps the best, the simplest, the truest definition of how they painted is Monet's famous remark to his chronicler Lila Cabot Perry:

> when you go out to paint, try to forget what objects you have before you, a tree, a house, and a field or whatever. Merely think here is a little square of blue, here an oblong of pink, here a streak of yellow, and paint it just as it looks to you, the exact colour and shape, until it gives your own naive impression of the scene before you.

There is as succinct a description of Impressionist practice as you will encounter: observe and record. Until we get to Seurat and neo-Impressionism it is inappropriate to speak of any unifying colour theory of Impressionism. It is true that Chevreul's *Principles of Harmony and Contrast of Colours* (published in 1839) was influential in a general way, but the principles of colour contrast that he proposed were only acknowledged in Impressionist paintings in the simplest terms: indeed, such use of colour had long been implicit in the practice of Delacroix and others.

Impressionist use of colour is quite straightforward. They rejected the prevailing academic notions of *chiaroscuro* – modelling in light and shade – and constructed their images in terms of pure colour, heightened in many cases with white, the technique known as *peinture claire*.

The most striking feature of many of these pictures is the bold juxtaposition of complementary colours – blue with orange, red with green, yellow with violet – all seen in Monet's dazzling *Regatta at Argenteuil* (Figure 7) of 1872. Chevreul had described how such complementary pairs mutually enhance each other by simultaneous contrast. It has to be said that the idea was anything but new; it goes all the way back to Alberti's colour chords and the 'friendship of colours'.

Renoir's *Boating on the Seine* (of 1879 or 1880) shows a vivid blue–orange pairing – and the pigments here are nearly all nineteenth century inventions: cobalt blue, chrome yellow, chrome orange, lemon yel-

Figure 7 Monet, *Regatta at Argenteuil.* (Musée d'Orsay, Paris.)

low and viridian, used almost unmixed, as they came from the tube.

The Impressionists were captivated by the purity and brilliance of the new pigments. Their avowed dislike of the old dark colours led to a principled rejection of black and the earth pigments that was more propaganda than fact. 'There is no black in nature' they cried, but black lingered on their palettes long after they claimed to have given it up.

Nevertheless, they sometimes went to extraordinary lengths to avoid using dull earth colours. In his *Gare Saint Lazare* (1877; Figure 8) Monet has mixed no less than seven high-toned pigments together to make the dark station canopy that could easily have been painted with brown and black. Throughout the 1870s he adopted this procedure, clearly intrigued by the almost imperceptible shimmer of powerful colours in the darks of his compositions.

The use of coloured shadows by the Impressionists is a well-known

Figure 8 Monet, *Gare Saint Lazare* (The National Gallery, London.)

characteristic of their work, and is something much more general than Monet's elaborate pigment mixtures. Goethe, Delacroix and others had long before noted violet shadows in yellow drapery and greenish shadows cast by a red sunset and had realised that these phenomena are caused by the physiological reaction of the human eye to powerful colours. 'Every decided colour', said Goethe, in his *Theory of Colours*, 'does a certain violence to the eye and forces it to opposition'. In other words, after looking at a strong colour for a while, everything becomes tinged with its complementary.

Coloured shadows in Impressionist paintings generally contained blue/violet tones as the complementary to yellow sunlight. Much fun was had at their expense, and the derisive term 'violettomania' was coined to describe their 'collective sickness' of painting 'people in violet woods' and 'purple-tinted corpses in a state of decay'.

Monet in his *Rouen Cathedral* series pursued the complementary coloured shadow to its ultimate point: almost abstract patterns of blue and orange, shimmering before us.

The use of colour by the Impressionists was essentially empirical and it is only when we reach Seurat and the neo-Impressionists that we have to consider a theoretical or scientific underpinning for their art.

Seurat

Seurat presents an enormous problem to students of art history and science alike. In the century since he painted, numerous scholars have accepted the premise, first disseminated by his friend the critic Félix Fénéon, that Seurat's technique of 'chromo-luminarism' imitated the very behaviour of light itself: that dots of pure spectral colour could re-form into the colours observed in nature with a brilliance and luminosity unobtainable with conventional pigment mixtures. Seurat, we were told, had based his method firmly on the scientific principles of such colour theorists as Chevreul and Ogden Rood.

In recent years a distinct reaction has set in to this mythology. We now know that Seurat misunderstood much of current colour theory and that, in many cases, the optical mixtures he so carefully calculated simply do not work, and they are certainly no more luminous than ordinary pigment mixtures.

This then is our problem. Scientifically speaking, there is a great deal less to Seurat than meets the eye: but his flawed and partial theories have given us some of the most haunting images in European art and all our reservations about his science must not blind us to that fact.

If we peer through the complex web of myth and pseudo-science that has been endlessly spun around his work, we can see that Seurat was, in fact, applying just two simple principles of contrast derived directly from Chevreul. Enhanced contrast of tone is seen very clearly in the light and dark haloes around the boys in the centre and at the right of the great *Baignade à Asnières* (p. 53) – painted in 1884 before Seurat's pointillism had begun, but retouched with some pointillist details in 1887. Next to the lit sides of the boys' bodies the water has been consciously darkened and next to the shadowed sides of their bodies it has been lightened.

The other contrast, that of complementary colours, is used everywhere in Seurat's optical mixtures, both for coloured shadows and for mutual enhancement of adjacent areas. Around many of his later pictures, he included a painted pointilliste border in which the dominant colour constantly changed and became the complementary of that part of the painting nearest to it – orange next to the blue sky, red next to green grass and so on. In his picture of *La Crotoy* of 1889 (now in Detroit) the original painted frame, with just such a colour scheme, still survives.

But obsessive pointillism killed spontaneity and careless optical mixtures actually killed colour as Signac, Seurat's fellow neo-Impressionist was later to admit. Writing in 1894, he said:

> Pointillage simply makes the surface of the paintings more lively, but it does not guarantee luminosity, intensity of colour or harmony. The complementary colours which are allies and enhance each other when juxtaposed, are enemies and destroy each other if mixed, even optically. A red and a green if juxtaposed enliven each other; but red dots and green dots make an aggregate which is grey and colourless.

As many have since observed, that greyness is palpably there, hovering over many neo-Impressionist paintings. To some it is part of their magic; to others it just adds a further dimension of unreality.

We have come a long way from the artificial colouring system of Cennino to the equally artificial system of Seurat. But I want to leave you with a simple image which, to me conjures up the whole charmed relationship between painters, their materials and their subjects. It is this small landscape painted in 1878 by Camille Pissarro on his own palette (Figure 9). The six colours he has used for his picture are all there around the edge, high-toned Impressionist colours out of which he has made something quiet, harmonious, timeless. No elaborate colour theories here, just a painter at work, observing and recording in pure colour – practising, as Paolo Pino, the sixteenth-century Venetian author called it, the 'true alchemy of painting'.

Figure 9 Pissaro, *Landscape Painted on his Palette.* (Sterling and Francine Clark Art Institute, Williamstown, Massachusetts.)

Further reading

Ackerman, J., 'On early Renaissance colour theory and practice.' In *Studies in Italian Art and Architecture, 15th-18th Centuries*, ed. H. A. Millon, pp. 11–40, Cambridge, MA, 1980.

Baxandall, M., *The Limewood Sculptors of Renaissance Germany*, New Haven, CN: Yale University Press, 1980.

Bomford, D., Brown, C., and Roy, A., *Art in the Making: Rembrandt*, London: National Gallery Publications, 1988.

Bomford, D., Dunkerton, J., Gordon, D., and Roy, A., *Art in the Making: Italian Painting before 1400*, London: National Gallery Publications, 1989.

Bomford, D., Kirby, J., Leighton, J., and Roy, A., *Art in the Making: Impressionism*, London: National Gallery Publications, 1990.

Dunkerton, J., Gordon, D., Forster, S., and Penny, N., *Giotto to Dürer: Early Renaissance painting in the National Gallery*, London: National Gallery Publications, 1991.

Gage, J., 'The technique of Seurat: a reappraisal', *The Art Bulletin*, **69**(3) (1987), 448–54.

Hall, M., *Colour and Meaning: Practice and Theory in Renaissance Painting*, Cambridge: Cambridge University Press, 1992.

Kemp, M., *Science and Art*, New Haven. CN: Yale University Press, 1990.

Lee, A., 'Seurat and science', *Art History*, **10**(2) (1987), 203–26.

2   Colour for the Painter

Bridget Riley

Introduction

For all of us, colour is experienced as something – that is to say, we always see it in the guise of a substance which can be called by a variety of names. For instance, pale, golden yellow may be the colour of hair, of corn, of certain fruit, of a precious metal, of some flowers, or woven into fabrics, or fall as patches of light. We usually see colour as the colour of *something* – it is not a natural thing to see colour simply as itself alone, unless, of course, we happen also to be painters. For painters, colour is not only all those things which we all see but also, most extraordinarily, the pigments spread out on the palette, and there, quite uniquely, they are simply and solely colour. This is the first important fact of the painter's art to be grasped.

These bright and shining pigments will not, however, continue to lie there on the palette as pristine colours in themselves but will be put to use – for the painter paints a picture, so the use of colour has to be conditioned by this function of picture making. The proper term for this use of colour by practitioners of the art is *pictorial colour*. For some artists, and most surprisingly it has only been a very small number, the necessity to come to some sort of understanding of pictorial colour has prompted the most passionate of enquiries and some of the greatest artistic adventures. Obviously what is perceived in the world about us is the primary experience of colour, and for the painter it nourishes and sustains, even if, like me, that painter today should be an abstract artist. Perceptual colour is our everyday experience of colour and, like Nature itself, it is a common condition. As long as we possess sight, these sensations are our constant companions, regardless of the degree to which we are aware of them. So the painter has two quite distinct systems of colour to deal with – one provided by nature, the other required by art – perceptual colour and pictorial colour. Both will be present and the painter's work depends upon the emphasis he places first upon the one and then upon the other.

Cézanne's remark in his letter to Bernard of 23 December 1904, that 'Light does not exist for the painter' refers to this duality. He has only the pigments on his palette and from those he has to fabricate, bring about any sensation of light that he wants in his picture. As it is with light, so it is with other visual and plastic sensations. For perceptual space the painter has to invent pictorial space. The same applies to our perceptions of form and weight, etc.: each sensation must be recast in pictorial terms. And if these are to 'work', as painters say, then together they must create a pictorial reality which is credible – so a painter has to find a way of uniting all the elements in a picture to make a whole.

The unity of colour: Titian, Monet and Caravaggio

In European painting, a remarkable development occurred in Venice at the beginning of the sixteenth century, which introduced an entirely new approach to colour. This was a unique event and one which did not take place in any other tradition of painting, not even in Persian or Chinese art, despite their great sophistication. It seems that it might not have happened at all without the insight of one great artist – Titian. Previously and in other traditions, painters had used what is termed *relational colour* in their pictures. This means thinking of colours as individual hues and placing them in the picture on a system of balances – a red against a blue, or a soft shade against a brilliant one, or one pale colour against another. Whatever the particular colour may be, it is treated as a separate item and given a specific role of its own to play in the organisation of a picture. Many great and beautiful works of art have been made using such a 'relational' approach to pictorial colour.

But it also opens up a problem – as one can see in Titian's early work. In *St Mark Enthroned and Other Saints* (1510, Santa Maria della Salute, Venice) parts of the figures and coloured draperies seem strongly dislocated and misplaced – there is a conflict between projected space and colour in the picture. The blue drapery which covers St Mark's knees is difficult to place – it is an oddly flat shape which has no form and seems to float somewhere above the column on which St Mark sits. *L'Assunta*, which Titian painted for the Frari church in Venice some eight years later in 1518, shows a very different approach. One would think that a scene describing a man sitting on the top of a column with a group of people standing on the floor below would be more readily convincing than a young woman borne aloft on clouds and surrounded by angels and cherubs. But this is not so – it is the 'imagined' scene, the invention, which is credible and this is entirely due to Titian's new solution to the colour problem. Here the colour, instead of conflicting with the projection of space, is meshed as one with it – they are more than inseparable – the colour actually describes the forms and their position in space.

Titian treats colour in the singular, as colour not as colours; he does not treat them as separate identities to be balanced one against another, but uses colour all together as one single element, which shapes and forms his painting. The colour sensation of *L'Assunta* is a warm rose, which runs through the entire composition. It turns towards brown in some of the clouds supporting the Virgin, takes a clearer rose in the cherubs, goes through oranges and browns towards yellows behind her, to be driven towards the strongest red hues in her robe and those of the two worshippers in the foreground below. Their reds, in turn, emerge from browns, magentas, dull red violets – countless variations on the rose theme. This whole scheme is laced with, and thrown into contrast from time to time by, the greens, blues and whites of other robes, the draperies and the sky. As this multi-hued colour moves through the painting, turning from shade to shade as it goes, it describes the great plane of the clouds, the bodies of the cherubs, the immensity of the celestial heavens, the limbs and the figures, and places them firmly in an invented spatial structure which is entirely credible. This extraordinary new way of *drawing* a picture, with *colour* rather than line, is the beginning of Titian's great innovation, the approach to colour which was to put European painting on a road that no other tradition had trod before, or for that matter, since.

At the other end of this tradition, which sprang from Titian's astounding insight into the unity of colour, stand the Impressionists and Post-Impressionists. Monet's approach came from a different corner, but it arrives nevertheless at the same point. His pictorial colour is unified by a perceptual '*enveloppe*', as he called it. This famous envelope is the bridge by which Monet can cross from the thing seen to the thing painted – to seize the very light and air, as it were, which envelopes a subject and to analyse precisely the qualities of colour which make it up. It is a way of accounting for the fact that in Nature too we can see the one 'colour' which defines the impression of the many. Each envelope provides a particular unity of colour and so each canvas in a series is able to present a subtly different sensation.

If you compare two paintings from the series he made in the valley of The Creuse in 1889, *La Petite Creuse* (Art Institute of Chicago), and *Petite Creuse, Sunlight* (Private Collection) you can see what Monet was able to achieve through his famous envelope. Both paintings show the motif from exactly the same point of view, but at slightly different times of day, under slightly different weather conditions, and at slightly different dates in that winter of 1889. All these small differences mean that the whole motif is distinguished by a different sensation of colour on each occasion, one that not only affects the actual local colours themselves, but that also produces different colours for the lights and the shadows, and in the reflections.

In the Chicago painting of *La Petite Creuse*, you see cool sunlight running along the distant hill tops, catching the sides of the valley and skimming over the rocks and ripples of the river as it flows towards you – a sparkle of the palest yellows. This unmistakable winter light is accompanied by cool, luminous shadows. High in key, they back up the colour of the light with transparent blues in the distance, gradually stepping up intensity as they advance until in the foreground, via the violets, they yield to the reds. In addition, the complexity of all the shadows, near and far, are compounded by reflected lights, which dematerialise the forms of rock, shrub and hillside. What remains are insubstantial wraiths of luminous colour, which weave their way through the painting.

In the later canvas *Petite Creuse, Sunlight*, the sun is stronger. Warmer shades of yellow, pink and even light magenta, flicker through it as the light follows virtually the same course as before. Not surprisingly, greens, rising from the palest turquoise to pure viridian, appear alongside the blues in the shadow sequences. The reds of the shaded rocks in the foreground are more pronounced here than in the other canvas. Traces of bright oranges and greens light up odd unexpected areas of the painting and culminate in the group of trees growing by the river bed.

There is a very revealing little incident attached to this painting. Monet was working on this series in the Creuse Valley for most of the late winter or early spring of 1889 (March to May). He had begun this canvas and then had to leave for Paris. When he returned he found the little oak tree, which stands aside from the others on the river bank, had put out new leaves and this change so upset the envelope of his painting that he felt unable to carry on with it.

Here is part of a letter to Alice Hoschedé (8 May 1989):

> ... I'm going to offer fifty francs to my landlord to see if I can have the oak
> tree's leaves removed, if I can't I'm done for since it appears in five
> paintings and plays a leading part in three, but I fear it won't do any good
> as he's an unfriendly old moneybags who's already tried to prevent access
> to one of his fields, and it was only because the priest intervened that I
> was able to carry on going there. Anyway, therein lies the only hope of
> salvation for these pictures.

The following day he wrote:

> I'm overjoyed, having unexpectedly been granted permission to remove
> the leaves from my fine oak tree! It was quite a business bringing
> sufficiently long ladders into the ravine. Anyway it's done now, two men
> having worked on it since yesterday. Isn't it the final straw to be finishing
> a winter landscape at this time of year ...   (Kendall, 1989, p. 132)

The problem of this little tree highlights the creative paradox which was to give Monet such anguish and frustration whilst at the same time being responsible for some of his most ravishing canvases. Monet, in order to re-create a unity of colour in his painting had, *sur le motif*, to feed his colour vision by acute observation. And it was only through resolving and re-resolving this creative paradox over and over again that Monet emerged as the great colourist we know – one in fact who found he *had* to invent a pictorial colour scheme as artificial in its own way as Titian's and, for the same reason that all artists adopt artifice, in order to come closer to a truth.

Although *L'Assunta* was a breakthrough and laid the cornerstone of

Western colour space, to see Titian developing this initial insight we have to look at the work of the 1550s. In his *Danae* (1553–4, Prado, Madrid; Figure 1), he clearly puts pictorial colour first and foremost, and by doing so achieves an unprecedented unity. The colour runs through the painting as though it were a single element, binding it all together. Look, for instance, at the colour of the hair, which follows the drawing of the far arm, breast and torso. This same hair colour turns into the colour of the contour of the near arm and side of the body. It is picked up and strengthened by the gold bracelet and reappears as the colour of the little dog. Traces of it also appear in the golden trimming on the magenta-coloured velvet curtains. The cool blues and greys of the painting are at their strongest in the sky and the grey-blue of the dark cloud. But this same grey-blue, which continues into the old servant's cap and blouse top, and is carried down into the shadows of the rumpled bed, is, in every case, a *plain* grey. It appears a *blue*-grey only through the contrast with the golden bronzes of the servant's skin and the yellow golds of the shadows on Danae's body.

Figure 1 Titian, *Danae.* (Museo del Prado, Madrid.)

A green, suggested by the landscape seen in the 'gaps' below the servant's arm and apron, re-appears on the other side of her body in the base of the stone column. This elusive green softens the juxtaposition of the magenta reds with the blues of sky and mountains.

The magenta reds of the drapery strike a chord of three tones: 'mid' on the left side in the foreground, 'deep' in the centre middle distance 'light' and 'brilliant' in the right foreground. The white bed linen is decorated by embroidered bands of magenta-pink in various strengths. This embroidery is a ruse by which Titian can take traces of the magenta from the draperies towards and even into the figure. The brilliant burst of golds, pinks and whites in the sky *is*, so to speak, the collective colour of Danae's body.

One might expect that with such an auspicious start, a tradition of colour painting would unfold quite smoothly. But it was not to be quite as simple as that. Titian himself was an overwhelmingly powerful figure: in addition to his achievements in the field of colour, his genius excelled on many artistic levels. Not only was he a hard act to follow in terms of stature, it was also extremely difficult to immediately penetrate such a complex body of work. The success of Caravaggio depended in part on his offering a manageable interpretation of Titian's approach by reducing the colour complexity to *chiaroscuro*, the dominance of tone over colour.

To be fair, on a superficial level Titian's colour painting can be read as a subtle gradation of light and dark values. Take, for example, the treatment of those magenta reds in *Danae*. If one isolates them from the other colours in the painting they begin to look rather like three different shades of the same hue. For someone trying hard to grasp precisely what principle lay behind Titian's way of painting, such passages could seem to offer not only a clue but also, very importantly, an intellectually comprehendable one. As a result, European painting was plunged into a heavy veil of chiaroscuro which was only sporadically lifted until the Impressionists arrived, who banished tonal painting and rinsed colour clean again.

In *The Burial of Christ* (1604, Vatican Museum, Rome; Figure 2) Caravaggio's new solution is fully formed. An impenetrable, almost uniform darkness, so pronounced as to be virtually black, accounts for the background. Strong white light falls on the group of figures carrying the corpse, throwing parts of them into sharp relief and others into deep

Figure 2 Caravaggio, *The Burial of Christ.* (Photo Vatican Museums.)

shadow and dominating the space and the modelling of the forms. The result is a dramatic contrast of light and dark, the very essence of chiaroscuro. Colour is reduced to a minor role, appearing only as a few isolated hues in the garments: brown for the tunic of the man lifting Christ's knees, dark blue for Mary's headress, red and blue for the robes of the man supporting the body. And that is all.

Caravaggio's influence was immense. It spread right across Europe, even into the most remote provincial schools of painting. The solution of chiaroscuro neatly circumscribed the problems raised by Titian's colour space. It also had the advantage of being highly conceptualised; that is to say, it introduced into painting a view of light as being separate from colour. This was in line with general assumptions; it could be described, understood from engraving, and practised almost immediately. Not surprisingly, it was a great success and became the mainstay of a good deal of European painting right through to the nineteenth century.

But Titian's original insight was simply too big to be ignored for long. Two very different artists took up the challenge and established a connection – one was Veronese and the other Rubens.

Veronese and Rubens

The stumbling block to establishing a colour tradition lay in the inescapable fact that no sound basis to, or guiding principle of, such a tradition could or can, even today, be found.

When Veronese arrived in Venice in 1555, Titian had just completed the *Danae*. He was the first painter to confront the difficulty of penetrating Titian's invention and his great contribution lay in finding a way of breaking down this barrier. In developing his own studio practice, he laid out clearly a step-by-step approach to the enigma of Titian's colour space. His solution is diametrically opposed to that of Caravaggio; it is focused unambiguously on colour, not on tone. He makes particular use of stuffs, decoration, architecture and objects, as agents or 'carriers' for the movement of his colour.

In the *Feast in the House of Levi* (1573, the Accademia, Venice), greys play an important but supporting role. Shifted towards a warm scale they appear in the architecture of the marble Loggia, which, with its golden lunettes, is set against a background of turquoise skies and pink buildings. Grey reappears as a neutral in the table cloth in the central opening of the Loggia.

The principal colour organisation is played out in all three openings. In each there is a strong red accent on the left, which kicks off the colour development. The most brilliant hues are reserved for the centre; the two sides are both comparatively subdued and virtually equal in weight.

In the central opening, Veronese has used vermillion against green in the foreground figure of Levi himself. From there, making use of cloaks and tunics, he pairs grey-blue with dull orange and turquoise-blue with magenta-pink in the figure of Christ. The passage finishes with echoes of vermillion with emerald green and a further repeat of very dark orange. Behind this group are two yellow figures of servants.

The colour movement in the left opening starts off with the red-orange figure in the corner, wearing a blue collar, who 'borrows' the red sleeves of the man in olive green next to him. After a sequence of grey-blue, magenta and golden ochre the initial chord, red-orange plus magenta, is repeated. A dark-brown figure seen against the light makes the transition to a green garment and an extension of gold and black-brown in the vessel and flask standing on the balustrade. Set against this group is a grey-and-yellow figure on the stairs in the foreground.

In the right opening there is again a strong vermillion, this time seen between the columns on the left, followed by red-orange, lime green, green-blue. The light magenta of the figure coming up the stairs relates to the darker magenta of the elderly man sitting in the centre behind the table. At the end, the man leaning forward and the servant with the out-stretched arm behind him are mirroring the constellation in the middle of the left opening: greyed lemon yellow against deep dark green. This whole chain of events is brought to a close by accents of grey-blue and red-orange.

Lastly the figures on the stairs and in the foreground: note the large man on the right, presumably in charge of the cellar, wears a magnificent striped costume, which gives Veronese a wonderful opportunity to use mauve magenta, grey, yellow ochre, red, green, greyed blue-turquoise and blacks together. It is the archetype of a certain sort of costume that Watteau was to paint over and over again later on. One can see a great deal of it in the Wallace Collection in London.

This strong chord of colour equals the green and pale magenta in the opposite group of Levi and the little black attendant. Between these two colour weights, in the nearest foreground, Veronese conjures up a truly magical demonstration of pictorial colour-perception. Throughout the painting a strong warm mauve sensation pervades the shadows and their reflected lights. But in the central opening, the colour of the tiles and of the dog on the floor below pick up this fugitive sensation and turn it into an actual hue.

Like Veronese, Rubens was essentially a 'do-er', a pragmatic artist, but he also had a strong analytical streak. He copied about thirty paintings by Titian, most of them during his last visit to the court of Philip IV in El Escorial in 1629 in Madrid.

If one compares Titian's *The Andrians*, c. 1520–2, in the Prado, Madrid, with Rubens' copy of it, now in Stockholm, one sees a very interesting shift in the colour. Rubens' painting is clearly based on red, yellow and blue, and their admixtures. Although Titian's certainly includes these primaries and tertiaries, he does not seem to have depended so essentially upon an organisation of these particular hues in shaping his colour.

There is another clue in Rubens' illustrations to *Six Books on Optics*, 1613, by Aquilonius. A diagram taken from one of them, shows Rubens demonstrating the relationship between the three primary colours and their derivatives in mixing. Obviously the development of his pictorial colour depends very much on his recognition of the physical properties of the pigments on his palette. At the same time red, yellow and blue are also to be found in the particular blending of the human skin. By seizing upon the correspondence between these two scales, Rubens seems to have found his colour key.

In the *Judgement of Paris* (1635–7, National Gallery, London; Figure 3), the first things that catch one's eye are the radiant bodies of the three goddesses, by far the lightest and brightest part of the painting. They are the prime movers of the myth and the origin of the colour in the picture. From their flesh tones, Rubens takes a compound of colour out into every corner of the canvas – here amplified to clear hues, there muted into faint shades. Pure reds, strong blues and golden yellows can be found in draperies, reflected lights, skies, distant views and shining surfaces. Subtle shades of violet, dull orange-browns and soft greens taken from body shades and suggested contours are carried out into earth, rocks and vegetation, up into clouds and crowns of trees: they continue in the two male figures, the animals and birds, and cross the foreground in the guise of tiny plants and the peacock's tail. The orchestration of this colour

Figure 3 Rubens, *Judgement of Paris.* (The National Gallery, London.)

movement in its grandeur and boldness is quintessentially Rubens – the plastic expression of an artist whose aim it was to master the best of the past.

Velázquez, Vermeer and Poussin

There are three other significant clues in the seventeenth century to the continuity of this particular approach to pictorial colour in the work of Velázquez, Vermeer and Poussin. It seems to be an important factor in colour's paradox that, being simultaneously singular and multiple, once treated as a whole it lends itself quite easily to a wide variety of expressions.

Van Gogh writing to Bernard says:

> Do you know a painter called Van der Meer? ... the palette of this curious
> painter consists of blue, lemon yellow, pearl grey, black and white. In his
> very few paintings there is in fact the whole richness of a complete palette;
> but the combination of lemon yellow, pale blue and pearl grey is as
> characteristic to him as the black, white, grey and pink are to Velázquez.
>
> (Complete Letters, B12, pp. 503–4)

Las Meninas (1656, Prado, Madrid; Figure 4), very clearly bears out the
truth of Van Gogh's observation. A vast canvas, over half the total pictorial area of which is taken up by a luminous darkness that envelopes the
farthest reaches of a huge room. In the foreground a group of figures
catch the light: courtiers surrounding a diminuitive Infanta. They are the
brilliant heart of the painting and the key to Velázquez's use of colour.

He treats black and white as colour, infinite shades of which, alongside
subtle greys, appear in the silks and velvets, linens and brocades, laces
and gauzes of the royal party. Pinks accompany the passage of light as it
falls across the figures. Starting with a strong accent in the suit of the
little boy below the casement of the window, they run up, in varying
strengths, through reflected lights, flushed skin, scarves, flowers and rib-
bons, to the palette of the painter in the background and end in the
Maltese cross on his breast.

Almost as though they were part of some visual amplifying system the
greys of the little group echo and reverberate in the modulations of walls
and ceiling above. Hovering between warm and cool, they generate the
impalpable grey space which provides the setting for the sharper contrasts below.

The secret of Velázquez's celebrated greys and blacks is not to be
found in any tonal concept such as chiaroscuro, but in his treatment of
them as true hues in themselves. For this recognition he was later much
loved by the French painters of the nineteenth century, and by Manet in
particular.

Van Gogh is right to compare Velázquez with Vermeer and to note the
characteristic difference in their colour. For Vermeer this is centred in
blue and yellow as it was also for Van Gogh himself. Blue being a darker
colour than yellow, which is light, there is the advantage of a 'quasi-tonal'
scale which springs directly out of the relative brilliance of these hues in

Figure 4 Velázquez, *Las Meninas.* (Museo del Prado, Madrid.)

the natural order. Through this, Vermeer was able to transform chiaroscuro into colour painting.

In *The Milkmaid* (*c.* 1658, Rijksmuseum, Amsterdam), yellow and blue in their purest chromatic state can be seen almost in the centre of the painting in the front of the maid's yellow bodice, where it catches the light, and in the top of her apron just below (Figure 5). Unlike the Infanta in *Las Meninas,* who leads the way *into* Velázquez's colour, the milkmaid draws together as contrasts the two colours which pervade the painting.

Figure 5 Vermeer, *The Milkmaid.* (Rijksmuseum-Stichting, Amsterdam.)

Lighter or darker, warmer or cooler yellow and blue account for the form, the space and the light, and are emphasised as the leading hues by the singular presence of the strong red in her skirt.

Take the yellow theme: from the bodice it lights up the inside of her cap as a luminous shadow, continues up the shadowed wall, including the brass object and the wicker basket, growing as it goes warmer and stronger until it reaches a dull orange in the top left corner.

The shadowed side of her bodice is also turned towards orange, likewise the shadowed side of her face and cap. It becomes stronger in the shadows under her forearms and turns into a heavy red-orange in the pitcher and vessel into which the milk is poured. It turns cooler and lighter again in the golden ochres of the bread and its basket.

The blue movement shows its deepest and strongest note in the shadowed apron and grows lighter in the table cover and the cloth lying across it – one a cooler slightly turquoise version, the other a warmer, more violet shade, respectively. It appears as both light and dark blue in the decorated jug and cover standing on the table, turns into the cool reflected light in the shadow below the window, and in the lightest and freshest of all possible inflections accounts for the colour of the daylight as it falls across the plain white wall behind.

Poussin is the last of the great intermediaries, neither Renaissance nor Modern, who contributes yet another aspect to the development of colour painting. A formalist, as at heart are Veronese and Rubens, he feels the need to organise his colour according to a precept. The practices of both the other artists are rooted in the practicalities of organising their studios. In the face of the bewildering diversity and sovereign impenetrability of Titian's legacy they are, in their different ways, great 'sorters-out'. Poussin, however, is working just that little bit later and so is in a position to raise a further question – that colour *can* be organised is one thing, but what should determine this organisation is yet another. In answer he came up with a curious reference to the Greek theory of music. There is a letter to Chantelou (24 November 1647) in which he talks about his way of composing a painting in relation to the musical 'modes' of antiquity.

Although art historians quite rightly doubted this comparison, it cannot be denied that Poussin organises his colour according to particular scales and harmonies. The red he uses for a Bacchanal, for instance, will be warm, bright and joyful, whereas the one he uses for a painting from the *Seven Sacrements* series will be cooler, deeper and possess altogether more ceremonial *gravitas*. Obviously his idea was that, depending upon the nature of his subject matter, he would shift his whole colour harmony. But the interesting thing is that Poussin does not carry this idea through *literally*. There is no system, no impeccable logic governing his colour. What Poussin is searching for in the spirit of music is a *method*: a method for arrangements in colour, for orchestration, for modulation, for chords, for the establishment of different keys or scales, for the composition of rhythms and movements. If so, the comparison to the musical 'modes' is a metaphor.

Some of the most beautiful paintings of dance ever made have come from his hand. In his *Adoration of the Golden Calf* (c. 1635, National Gallery, London), Poussin takes red, blue and golden yellow at their purest for his primary chord and places it in the robes of the three figures low down, in the bottom right foreground. These three hues then change key and, relieved by the addition of white, repeat in the figures dancing round the Golden Calf on the left. Modulated in different strengths and tones they move back through the crowds in the middle distance on the right. Traces recur as muted shades in the distant mountains and the banks of clouds. The idol carries up the golden yellow and pitches it against the rose pinks and blues of the sky to orchestrate the climax of Poussin's colour movement.

This is a new development in European painting. It is the first genuine 'method', a way of thinking which has arisen in response to a newly felt need. With Poussin the abstract substructure of picture making comes very close to the surface. It is for these reasons that he has been called a painters' painter – and it may also be the rigorous and uncompromising way in which he dealt with his problems that made him such an important artist to Cézanne.

Cézanne and Delacroix

Poussin's colour is pictorial colour in an absolute sense and this is in part what Cézanne is referring to when he speaks of 'doing Poussin over again from Nature'. For Cézanne is an artist who works *sur le motif* and he is caught in the same creative paradox as Monet in the conflict between pictorial colour (that which has to happen on the canvas and which has to obey the laws of picture making) and perceptual colour (that which he

Figure 6 Cézanne, *Great Pine and Red Earth.* (Hermitage Museum, St Petersburg/The Tate Gallery, London.)

observes out there in Nature and is, for him, the precious source of his 'sensation').

The one thing for which Cézanne feels a need is a method – 'avoir une belle formule' as he puts it in answer to the question 'What constitutes human happiness?'. In painting Poussin seemed to possess this.

It is the origin of modulation in preference to modelling as a new way of 'drawing'. In *Great Pine and Red Earth* (c. 1885–7, Hermitage, St Petersburg; Figure 6), Cézanne establishes the picture plane through the tree trunk and its branches. Through his colour modulations he 'draws' a new pictorial space – one created in response to the sensations the motif has aroused in him. In Cézanne the implications of Venetian colour-space are taken to quite astounding lengths. In literal terms the tree would be growing at the junction of middle distance and foreground; in sensation it is the nearest thing to you. Passages of landscape seen through the branches are detached from their topographical location and re-formed into one continuous but dislocated plane identical with that presented by the tree trunk.

Colour modulation is the agency for this transformation. Blues, pinks, yellows, greens and luminous greys in the highest key are contrasted with related hues at their most sombre. This colour compound, generated by the tree trunk and landscape, gives the key for the mid-tone modulations which carry the blues and greens, reds and bronzes through the whole painting. The colour is so wonderfully integrated as a cohesive whole that the entire surface of the canvas gives off an extraordinary soft pearly light, an exquisitely judged result of colour interaction. This was much prized by Cézanne; he wrote (23 October 1866) to Pissarro 'You are absolutely right to talk about the grey which alone rules in nature but it is frightfully difficult to achieve'. One finds light and dark, warm and cold, soft and sharp sensations of colour orchestrated in spatial rhythms and movements. The substance of form is de-materialised. Bushes, shrubs and undergrowth disintegrate – the blue of the sky comes down into the foreground as reflected light; earthy orange-yellows appear in the crown of trees; reds pull up a distant house, push down a patch of foreground. A whole new pictorial order is created.

The perception of forms thus 'drawn' and the spaces established between those forms, although generated by the sensation Cézanne experiences before Nature, finally emerge firmly built on the canvas as purely plastic inventions. The method which makes it possible to realise these inventions, to translate sensation into pictorial reality, is comparable to the one which he perceived in Poussin – a way of thinking which would allow him to orchestrate the 'harmonie générale' of nature into an 'art as durable as that of the Museums'.

The art of the museums also fascinated Delacroix. In fact these two artists, besides one being a hero for the other (Cézanne made studies for an apotheosis of Delacroix being gathered to the Heavens of Great Art), shared a veritable pantheon of colourists from the past from whom they drew strength and inspiration. 'The Great Venetians', as Cézanne called them, tower over all for both artists. Both followed the same route in rediscovering Titian's colour-space – in the beginning by being attracted to Rubens, and then by recognising Veronese as *the* principle interpreter. Delacroix is the first link in the fabled 'golden chain' forged in the nineteenth century by the great French painters. He copies, he observes, he learns and he reflects. For all the image of Delacroix as the Romantic artist *par excellence*, working in the heat of passion, his great admirer Beaudelaire saw him as 'sang-froid animé': 'Delacroix was passionately in love with passion and coldly determined to seek the means of expressing passion in the most visible manner'.

He developed a habit of going straight to the Louvre whenever he ran into difficulties in the studio to see what the others had done. It is almost as though he held a running dialogue with the past and if he couldn't understand from the work itself he then looked to Nature as 'the dictionary' in which he could find the missing word, so to speak, to illuminate the painting which he couldn't read. What finally emerged is the birth of 'peinture claire', as Monet hailed the wall paintings in St Sulpice: the liberation of colour from chiaroscuro. But long before St Sulpice, which was completed towards the end of Delacroix's life, he had paid a visit to Tangier. What he saw there made a deep impression on him: Mediterranean light, with its wealth of reflections, its luminous shadows and, above all, the focus it turned on the play of colour in this strange exotic world in which he found himself.

In *The Women of Algiers* (1834, Louvre, Paris), one sees a large patch of reflected light falling on the tiled wall of the interior in the background and luminous shadows on the head and body of the girl sitting in the centre foreground. They are, respectively, a fresh light blue-turquoise and a softly glowing creamy gold: both colours in a middle tone and together making a subtle, near complementary, contrast in the heart of the painting. They offer a ravishing clue to Delacroix's colour thinking. Reflected light and luminous shadows tend to appear as mid-tone colours – 'demiteints' as he called them – and he noted this pitch as being ideal for the perception of colour. It is precisely in those reflected lights and luminous shadows which are fugitively present in every part of the canvas that the colour key is to be found. One can well understand that Renoir, who as a young man was nicknamed 'Delacroix blond', loved this canvas as 'the most beautiful painting in the world'.

It is revealing to take a close look at the white embroidered blouse worn by the black-haired girl half-sitting to the right of centre. This blouse has a tiny sprigged design of pink and green: a little flower and a few leaves. Where the strong light falls across the front of the girl both pink and green appear dark and almost indistinguishable in contrast to the white. In the area of shade cast by her arm, the design shows fresh clear colours at the same mid-tone pitch as the luminous yellowish grey of her shadowed blouse. That is to say, when the tonal contrast is at its strongest the colour is diminished; conversely when the tones are equal the colour contrast becomes clear.

Delacroix left little to chance; his work was meticulously planned, often long in advance. In his last great piece *The Expulsion of Heliodorus from the Temple* in St Sulpice, for instance, one can see in the foreground the temple treasure box, which has fallen open after it has been dropped by Heliodorus. Delacroix dictated the entire colour scheme of this detail to Pierre Andrieu, his assistant, who wrote it out in his notebook in March 1852, *eight* years before the completion of the painting.

Today we know Delacroix's way of thinking through his journals and notes but these were not published until 1893, thirty years after his death. His intellectual influence was largely exerted through his friend, Charles Blanc, the founding editor of the magazine *Gazette des Beaux Arts*, who popularised his work and his 'theories'. Being also a close friend of Chevreul, Blanc may have presented these in an overly dogmatic, quasi-scientific manner, but in fact both Delacroix and Chevreul arrived independently at similar conclusions in understanding the structure of colour contrast. Years before Chevreul became director of the Gobelin Workshops and began to develop a professional interest in colour, Delacroix had noted the effect of simultaneous contrast in the collar of a servant standing on the left-hand side of Veronese's *The Marriage of Cana* (1562–3) in the Louvre.

The Impressionists and Post-Impressionists drew on all three sources. In Cézanne's studio in Aix one can still see Blanc's books lying on a shelf. The main ideas and principles are very few and rather simple – it is only in practical application that the complexity of colour's intricate nature, dependent as it always is upon relationships, becomes apparent.

Seurat, Monet, Matisse and Picasso

To Seurat it may have seemed only a small step to push further the latent scientific aspect of this perceptual heritage and to secure for colour a firm and certain basis, and to find in the century of great discoveries a fitting place for the art of painting in the very forefront of contemporary science.

He first identified and divided up the composite nature of colour in the thing perceived. These portions then had to be accounted for through specific hues and to be located in appropriate quantities of little separate touches of paint, a technique which became known as *pointillisme*. The critic Fénéon recalls Seurat making the following divisions:

1. *local colour*: 'that is the colour that the surface would have in white light (appreciably the colour of the object seen from up close)'.
2. *directly reflected light*: 'the portion of coloured light that is reflected unaltered from its surface (this will probably be *solar orange*)'.
3. *indirectly reflected light*: 'the feeble portion of coloured light which penetrates below the surface and which is reflected after modification by partial absorption'.
4. *colour reflections projected by neighbouring objects.*
5. *ambient complementary colours.*

This schematic translation of colour perception into pictorial colour was the cornerstone of 'ma méthode'. The list itself was actually written out at the time of *La Grand Jatte*, his great masterpiece, but the ideas it set forth had been gradually evolving for quite some time.

In the *Bathers* (*Une Baignade à Asnières*, 1884, National Gallery,

London; Figure 7), Seurat's first great painting made in one year and completed by the time he was twenty-five, 'ma méthode' is not yet perfected and in place. Nevertheless, just because of that, one can see his way of thinking more clearly there than in a fully conceptualised piece like *La Grande Jatte*.

The first thing one notices about the *Bathers* when one catches sight of it in the National Gallery is the cool fresh light which radiates from the whole surface, as though the sensation of pure daylight itself had been caught and pinned down there for us to take part in.

One sees, of course, the figures on the bank, in the river, the trees, the sky and distant buildings, but one sees them all bathed in this extraordinary soft pearly light, which seems to be both clear and at the same time full of colour.

Figure 7 Seurat, *Une Baignade à Asnières.* (The National Gallery, London.)

The painting is to a considerable extent a great Impressionist painting made not *sur le motif*, but from many preparatory studies in the studio. The principal colours which are responsible for this beautiful radiance are orange and green, blue and violet. They are mostly present at a high or mid-tone pitch. They are two pairs of complementaries which tend on the small scale of Seurat's brush marks to give off coloured greys. This contributes to the pearly look of the painting. The optical mixture of the whole could be described as layered: there is the interaction between the various hues as they are put down on the canvas, a relatively simple 'physical' interaction; and between the induced colours themselves and these painted hues a kind of secondary more complex interaction takes place. Most importantly these two sorts of optical mixture are *not* successive but simultaneous and cumulative, and together they generate the wonderful light of the painting.

Seurat works into this body of colour in several very curious and what may seem sometimes almost incompatible ways. Today the *Bathers* might perhaps be described as a transitional or a *proto-pointilliste* work, and at first sight Signac, Seurat's eager disciple, criticised its inconsistencies quite severely. There is a great deal of plain local colour for instance: skin where it catches the light is to a large extent pink; trousers, hats and boots are brown; panama hats are straw coloured; and considerable portions of the white articles, such as clothes, towels or sails, remain white.

At the same time, Seurat deliberately painted around some of these plainly stated forms according to the principle of irradiation, which has been known to painters since Leonardo observed and noted it. It says that if a light form is seen against a dark ground the contrast intensifies both light and dark edges. In the lying figure in the foreground one can see this principle being applied in a classic manner: along the upper edge of the dark trousers Seurat has introduced into his divisionist treatment of the grassy bank touches of pale colour, and where this silhouette changes from dark to light in the white jacket, little strokes of darker colour have been added to the grass alongside the white jacket. A similar treatment is used for the auburn-haired boy sitting on the bank in the centre: the front of his shoulder and forearm turn an exceedingly pale cream against a river dramatically darkened by divisionist technique into a multi-hued deep blue. This is a very powerful contrast and one which could be expected to rupture the unity of the painting, but, astonishingly, it does not. The plastic coherence which Seurat has built is so strong and resilient that it can absorb even this disruption. Conversely, the boy's entire back and shadowed face are darkened against a lightened river. The same treatment of the irradiation phenomenon is repeated unequivocally in the contours of the two bathers in the river. This way of contrast thinking is taken from tone right into colour in the figure of the boy sitting in the straw hat in the middle distance. There, pure local colour is virtually laid aside and divisionism unifies the handling. The shadowed back is turned towards violet; the hat and the singlet strongly so, and at the demarcation line Seurat has added many little dots and dashes of yellow-green to the grass. This same singlet shows two other remarkable instances. Round the arm-hole, touches of turquoise-green are juxtaposed with the reddish orange shadow under the arm, and then green strokes turn up again, even more pronounced, at the edge of the magenta sash around the waist to which Seurat has added rosy red. Finally, more difficult to see but unmistakably there none the less, one finds along the bottom of the dark trousers deep blues next to strong oranges pulled out of the browny mauve shades of the rug on which the boy is sitting.

In the remaining few years of his short life, Seurat went on to try and iron out all the irregularities in his *méthode* and in doing so he created some of the most mysterious, beautiful and extraordinary pictorial inventions ever made. But the conflict between his achievement and his aim deepened. The more inflexible Seurat's method became, the less it could fulfil the role he had in mind for it. Each and every painting made is unavoidably a unique fabric of relationships. Change one thing and everything is changed – this complete dependence on the *particular* is not reconcileable with scientific ambition. But it is the very bedrock of creative speculation – the very fact that colour *can* only be determined through relationship prompted the enormous range of invention which its resources supported in the nineteenth and early twentieth century.

Monet in his great 1899 series *The Japanese Bridge*, which he painted in his garden in Giverny, is celebrating precisely this essential truth of colour. To Georges Clemenceau, he said: 'I am simply expending my efforts upon a maximum of appearances in close correlation with unknown realities. When one is on the plane of concordant appearances one cannot be far from reality, or, at least from what we can know of it.'

Like a scientist, Monet has a method, but not a scientific one. He takes the same viewpoint every time. In the foreground, where each particular colour envelope is clearly set forth, lies the stretch of water with its lily pads and reflections; the arc of the Japanese bridge above states the middle distance, cuts short the recession and helps to flatten the picture plane. The background of trees, shrubs and rushes frame and contain the little pond as the centre of attention.

The changing light, its shadows and reflections become the principal factors in determining which envelope Monet will select. The tonal structure plays a very interesting and surprising role, almost photographic in its verisimilitude it underpins with extreme sobriety the wonderful boldness of Monet's colour.

In *Water Lilies and the Japanese Bridge* (The Art Museum, Princeton; Figure 8), the colour chosen for the light, a pale fresh yellow can easily be distinguished in the lily pads as they lie in the early morning sun, and conversely they also show the contrast colour, a light blue-violet, in the shade. This 'pair', which precisely complement one another, is then echoed up through the rushes, through the little bridge, through the trees, saturating the entire image with their resonance.

Against these pale yellows and blue-violets, Monet threads, on the pretext of reflections in the water, a second pair of contrasts: the strong dark greens and reds which originate in a few traces tucked in amongst recesses and crevices in the upper wall of foliage. And yet a third pair, pink magentas and clear yellow greens, found in the lily flowers, buds and pads, in the shade or sunlight, compound the intricacies of the colour structure. In addition, reflected lights allow cool emerald greens to be pulled into play: they illuminate some shadows in the trees and appear fugitively in the rushes and water plants in the pond.

In *The Water Lily Pond* (National Gallery, London), the lily pond again sets the key; it is mid-morning, the colour of the light has taken on a more dominant shade of yellow, which now washes a larger part of the motif. The shadows are proportionally reduced and have turned into a few traces of an even lighter blue-violet. The reflected lights are again green, but stronger, and with a more pronounced presence. The reds and

Figure 8 Monet, *Water Lilies and the Japanese Bridge.* (The Art Museum, Princeton.)

greens reflected in the water are lighter in tone and both are much warmer. As a result of this shift in his envelope the whole sensation is entirely different.

In *Bridge over Water Lilies* (Metropolitan Museum of Art, New York), it has changed again. It is now noon, the sunlight is bleached to a creamy white and flattens the forms throughout. The shadows have faded into the palest bluish grey. Both appear to be virtually colourless and they are brought so close together tonally that they can hardly be differentiated.

If in this painting the motif is bathed in light, in *Water Lilies: Harmony in Green* (Musée D'Orsay, Paris), it is plunged into shadow and many staple points in the familiar tonal and colour structure are consequently turned inside out. There is almost *no* contrast: what comes nearest to such a relationship takes place when the two opposite ends of the magnificent green harmony are brought into proximity as yellow greens and golds, and as turquoises and blues. This happens through the agency of the trees in the background caught by the afternoon light. They appear as huge cascades of glorious golden yellows and yellow greens. Adjacent shadow passages of dark greens and turquoises throw these bright colours into relief, and in amongst these shadows go traces of blue as reflected lights. In this painting it is the background rather than the foreground that lays out Monet's envelope.

The last painting in this little group taken from the series is *The Japanese Footbridge and the Water Lily Pool, Giverny* (Collection of Mr and Mrs David T. Schiff). It is early evening, the light and shade form a dramatic tonal contrast, the 'action' has returned to the lily pond and once again the water plants play a central role: yellows, in a wealth of hues, for those in the evening light; clear blue-violets for those in the shadow. In between these two zones of lily pads lie astounding reflections in the water. These, through the complexity of perceived after-images in Nature, are strong reds.

Two interesting contradictions are at work in Monet. First, although each work was clearly conceived at a specific time of day, what we experience in looking at a finished canvas is a serene sense of timelessness. Secondly, despite the fact that each of these paintings is a beautiful and independent work of art in its own right, Monet's series method was indispensable in their creation – through that he was able to isolate, if we can use such a word about a group of relationships, a unique instance of colour's sensation from the myriads of such sensations. Transformed into pictorial colour a quintessence of this sensation could then be expressed.

If the precise character of such sensations could be given by colour in a painting then it is only a short step to speculate, as young Matisse appears to have done, that this insight could be reversed and that pictorial colour could become the vehicle of sensation. *The Open Window in Collioure* (1905, Collection of Mrs John Hay Whitney) shows pictorial colour as absolute. Matisse does not look for an equation to perceptual colour, instead he uses a mid-tone turquoise for the shadowed half of his interior and a magenta of the same tonal pitch for the other half where the light falls. He takes this pair of colours, the main contrast of the painting, out to the radiant sea-scape beyond the little balcony.

Palest turquoise accompanies magenta in the distant sky and sea; nearer and deeper in tone the two colours appear as the hulls and masts in the line of bobbing boats; light magenta and white are used for the glittering shallows along the foreshore. The same two colours, but much deeper and stronger, recur in the area above the open window seen in counter-light. Between the distant view and the interior Matisse inserts a filigree frame of secondary contrasts: low-keyed reds, greens and yellow greens as separate touches, insulated by surrounding whites, for the leaves and flowers of the climbing plant growing up the trellis and over the balcony.

The darkest tones, to punctuate the whites, are reserved for quite large blotches of deep blue-violet placed low down in the shade of the balcony and for big, bold marks of this same colour and black in the window reflections of the open casement.

Matisse prefers the word 'émotion' to sensation in describing his expressive aims. Here we are up against the difficulties of translation which can be resolved only by looking at what he did.

In 1911, he painted *The Red Studio* (Museum of Modern Art, New York; Figure 9). A single colour envelopes the whole volume of the studio and accounts for the walls, the floor and nearly all the furniture. Although it is an uncompromising red it is pitched at a tonal level that one immediately recognises as somewhat familiar; in fact Matisse originally intended to paint the studio the blue-grey that it actually looked like but once on the canvas the sensation this colour gave did not match up to his 'émotion' – to what he felt when standing in the room. So he changed it to red for, as he said, no reason that he could explain. Despite such abstraction, the tonal pitch of this red helps to express a particular soft even light which fills this interior.

The curtained window on the extreme left does two things: in its brilliance as nearly white it lends the red a sense of darkness and, with its cool shadows of transparent green, it seals off this enclosed space from the outside world. This near white is picked up in the bright glare of a number of objects around the studio, particularly the basket-work chair

Figure 9 Henri Matisse, *The Red Studio.* Issy-les-Moulineaux (1911) Oil on canvas, 71¼ in. × 7 ft 2¼ in. The Museum of Modern Art, New York. Mrs Simon Guggenheim Fund © Succession H. Matisse/DACS 1997.

on the extreme right, directly facing the window. The others, be they the decorated plate, the round dial of the clock-face, the plaster figure on the modelling stand, the ornaments on the chest of drawers, the painted areas in the canvases, the trailing arabesques of the plant in its vase, or even the shine in the wine glass standing on the table, *all* combine not only to maintain the prevailing sense of darkness but also to establish red and white as the principal contrast of the picture. Matisse modifies one side of this contrast by the other. A scale of tones is buried in a gradation of hues: high-keyed pink, for a quite dazzling quality of reflected light, in the big painting leaning against the wall by the window; deeper shades of pink and light rosy reds for softer reflected lights in the other paintings hanging on the wall in the background accompanied to a lesser extent by the yellows, greens and blues – all mostly confined to patches of colour in the canvases and decorative patterning on ornaments. The dark accents are extremely limited, deep greens in the leaves of the plant in the foreground and strong blues in the background.

In *The Pink Studio* (1911, Pushkin Museum, Moscow), painted a few months earlier in the same year, the approach is in many respects the opposite. The feeling, the *émotion* expressed by the colour is also one of a luminous interior but it is altogether lighter, fresher and airier. Matisse divides his principal colour into two shades: a pinky lilac, pale and cool, for the studio wall seen opposite in counter-light and occupying the upper half of the picture; and a rosy red, warm mid-tone, for the flattened plane of the floor below.

The cool feeling is developed by the whites concentrated towards the centre of the canvas in painted areas of pictures hanging on the wall, sheets of white drawing paper, and in the sunlit exterior, which seems to press forward through the window panes, and more emphatically in a light, fresh turquoise for the screen standing in the centre in front of the window – traces of this same colour paired with whites can be found in the window frame, the large statue and the background of a canvas leaning against the wall nearby. Although this colour provides a subtle contrast, the dominant one is to be found in the cold grey-blue of the darkly patterned cloth thrown over the turquoise screen in front of the window. The warm side of this pink harmony is taken up and developed by dull yellows, red-browns and olive greens shown mainly together. Various modulations of these warm colours circulate around the picture in various pieces of studio furniture and equipment (that is to say, in the carpet, the statuette, and the palette), in the paintings on the wall and in the leafy trees seen through the window.

Matisse and Picasso are two such different artists that they virtually seem to complement each other. Picasso did not make colour as central to his work as Matisse did. Nevertheless, after the Cubist adventure, his quick and agile mind instantly grasped the achievements of modern colour painting.

In *The Three Musicians* (1921, Museum of Modern Art, New York; Figure 10) we can see his handling of colour space in a Post-Cubist way. Maurice Raynal wrote of this painting that it is rather like 'a magnificent shop window of Cubist inventions and discoveries'. This is true but there is also a distinct feeling of Spanish darkness, the special light of

Figure 10 Pablo Picasso, *The Three Musicians.* Fontainebleau (summer 1921) Oil on canvas, 6 ft 7 in. × 7 ft 3¾ in. The Museum of Modern Art, New York. Mrs Simon Guggenheim Fund © Succession Picasso/DACS 1997.

Velázquez. The cubist planes are used as colours denoting positions in space which advance and recede and collectively build an airy volume in the centre of the painting. Black, white and browns laced with red, yellow and blue shift and slot dislocated parts of the three figures together in a syncopated rhythm. Browns are at their lightest for the floor of the stage on which the musicians play; deeper browns for the wings and backdrop are turned into a luminous space by the blacks moving through the composition from the shadowy figure to the left of the stage. More browns infiltrate the group, becoming gradually warmer and lighter as they advance in sharply angular planes, culminating in a red-brown for the flattened table top in front. A big blue shape, in contrast to the browns, glides between them acting as an elusive surrogate for a leg here, a reflected light there, a mask, a shadowed shirt and even parts of the ashtray on the table.

Warm yellows and reds, from the harlequin costume of the musician playing the bright yellow guitar in the centre, raise the tonal key of the group and bridge the interval to the whites. These, as planes of light, are interspersed among the figures as the sudden illuminations which flood and flicker over various features, music sheets or instruments, hats, arms and legs in a theatrical low light.

We are now almost at the present time and no more of any significance has been added to this great body of work. It is unique to European painting and remains so. The conditions are pictorial and the insights from Titian to Matisse originate in the recognition of this fact.

This does not mean, however, that the story has come to an end. There is certainly a pause, but too much has been invested by too many brilliant minds to be ignored and, though the spirit of artistic enquiry may sometimes sleep, it does not die. This wonderful adventure, which might be called 'The Tale of True Colour', has taken many twists and turns, fallen into disuse and neglect, has been re-discovered and re-invented by such very different painters working in specific centres in Italy, the Netherlands, Spain and France at very different periods in time.

Colour's potential is enormous in the hands of the painter and those relatively few artists from whose *oeuvres* I have selected only one or two paintings will have shown the reason for this richness. It is as though what might be considered colour's weakness from some points of view is for the painter its strength: just because there is *no* guiding principle, *no* firm conceptual basis on which a tradition of colour painting can be reliably founded, this means that each individual artistic sensibility has a chance to discover a unique means of expression.

Further reading

Baudelaire, *Selected Writings on Art and Artists*, Cambridge: Cambridge University Press, 1981.
Blunt, A., *Nicholas Poussin*, London: Phaidon Press, 1967.
Clemenceau, G., *Claude Monet. Les Nymphéas*, Paris: Librairie Plon, 1928.

Doran, P. M. (ed.), *Conversations avec Cézanne*, Paris: Macula/Pierre Brochet, 1978.

Elderfield, J., *Matisse in the Collection of the Museum of Modern Art*, New York: Museum of Modern Art, 1978.

Homer, W. I., *Seurat and the Science of Painting*, Cambridge, MA: Harvard University Press, 1964.

Kendall, R. (ed.), *Monet by Himself*, London: Macdonald Orbis, 1989.

Pablo Picasso: A Retrospective, Exhibition Catalogue of the Museum of Modern Art, New York, 1980.

Rewald, J. (ed.), *Paul Cézanne, Letters*, 4th edn, Oxford: Bruno Cassirer, 1976.

Riley, B. *Dialogues on Art*, London: Zwemmer, 1995.

The Complete Letters of Vincent Van Gogh, 3 vols., London: Thames and Hudson, 1978.

3   Light and Colour

Malcolm Longair

Introduction

Light and colour can mean very different things to different people. I breathed a sigh of relief when I learned that my assignment was to provide the physicist's view of light and colour because, to me, every other aspect of light and colour is quite horribly complicated. I will leave it to the other contributors to this volume to discuss the aesthetics, the art, the physiology, the psychology, the philosophy and so on. What is left for me?

The understanding of the physical nature of light and colour is a wonderful story and I will re-tell it from a historical point of view. This approach is not just fun but is also revealing of the real process of scientific discovery as opposed to the cut-and-dried version which appears in textbooks. It also reveals what I have found through experience: the aspects of the physics which were most difficult historically are also the bits that cause the greatest trouble to students and to non-scientific audiences.

My intention is therefore to describe some of the classic experiments of the last 400 years that have led to our present understanding of the physics of light and colour. It is also an opportunity to illustrate some of the more remarkable optical effects of light and colour that occur in Nature. At the end, I present the physicist's current view of the nature of light, and it acts as a beautiful exemplar of the type of issue which underlies a great deal of modern physics and the problem of visualisation of how nature actually behaves. I have to issue the usual disclaimer on such occasions that what I will present is the physicist's view of history rather than that of the historian or philosopher of science. This means two things. First, the history is written with hindsight and so the heroes are those who got the right answer. Secondly, I have borrowed quite unscrupulously from the history of science and re-written it for my own purposes.

It turns out that many of the great pioneers of the subject dealt with the two separate questions implicit in the title of this chapter – the understanding of the *nature of light* itself and the *nature of colour*. Although closely related, they are quite distinct intellectually, as we will find.

The bending of light

I begin my story with the discovery of the law of refraction by Willebrord van Snel van Royen in 1621. The law appears in Christiaan Huygens' great *Treatise on Light* of 1690 and describes the simple trigonometric relation between the angle of incidence a_1 and the angle of refraction a_2 of a beam of light on passing from one medium to another (Figure 1):

$$n_1 \sin a_1 = n_2 \sin a_2$$

where n_1 and n_2 are constants appropriate to the media through which the light ray is travelling and are known as their refractive indices. The

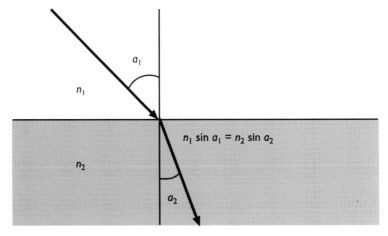

Figure 1 Illustrating Snell's law of refraction in which the angle of incidence of a light ray a_1 is related to the angle of refraction a_2 by the simple trigonometric relation $n_1 \sin a_1 = n_2 \sin a_2$, where n_1 and n_2 are the refractive indices of medium 1 (above) and medium 2 (below).

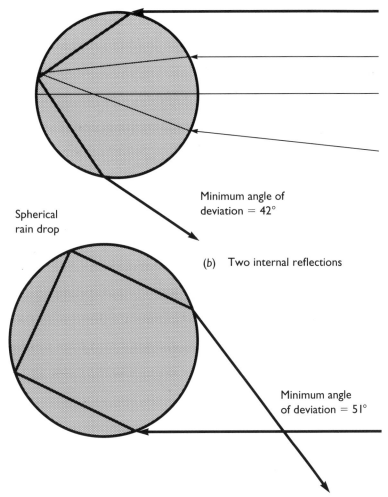

(*a*) One internal reflection

Spherical
rain drop

Minimum angle of
deviation = 42°

(*b*) Two internal reflections

Minimum angle
of deviation = 51°

Figure 2 Illustrating the refraction of light rays inside spherical raindrops.
(*a*) The case in which there is one internal reflection inside the raindrop.
The light rays are concentrated towards an angle of 42° to the line of
sight, which is the minimum angle of deviation from the direction of
incidence. (*b*) The case in which there are two internal reflections. The
light rays are concentrated towards an angle of 51°.

same law was discovered independently by Descartes and in France the
law is known as Descartes' law.

Almost immediately, there were two remarkable developments from
this simple law. The first historically was Descartes' explanation of the
origin of rainbows which appeared in his volume *Les Météores* of 1637. At

that time, the composition of light was not understood and so Descartes' theory described the production of a white rainbow. Nonetheless it contained all the essential ingredients of the modern theory.

Descartes asked what happens when light is refracted on passing through a spherical raindrop. The answer is that the deviation of light on entering and leaving the raindrop is exactly as described by Snell's law but in addition the light ray is internally reflected and so passes out of the drop on the nearside as shown in Figure 2(a). The interesting result of Descartes' calculations was that, when the rays are internally reflected once, as shown in the diagram, there is a minimum angle of deviation of 138° which corresponds to an angle of 42° with respect to the direction of incidence of the light rays. By tracing how light rays are refracted inside spherical droplets, Descartes showed that there is a bunching up of the light rays towards the angle of 42° – what we would now call a caustic surface. Therefore, we would expect the sky to be brighter just inside the rainbow as compared with outside it. If there is a shower of rain on a sunny day, and we look in the direction opposite to the sun, we expect to observe a concentration of light in all directions at an angle of 42° with respect to the line from the sun through our heads. Notice that it does not matter how far away the water droplets are from the observer. The only important point is that there is a preferred angle between the direction of the sun and the observer. Descartes also realised that, if there are two internal reflections instead of one, there will be another preferred direction, this time at 51° to the line from the sun to the observer (Figure 2(b)), which results in the appearance of a double rainbow. The essence of this argument is entirely correct but as yet there is no colour in the story – that had to await the experiments of Isaac Newton.

The second remarkable analysis was carried out by Pierre de Fermat, in 1657, who gave an alternative way of viewing the phenomenon of refraction. He enunciated his *Principle of Least Time*, according to which the path of the light ray on passing through different refracting media is such as to minimise the time it takes the ray to travel between the points A and B (Figure 3). According to Fermat's view, light travels more slowly in the denser medium, in fact just by the ratio of n_1 to n_2, and so we can work out which route takes the shortest time. I have drawn a number of possible paths in Figure 3 and it is a simple calculation to show that,

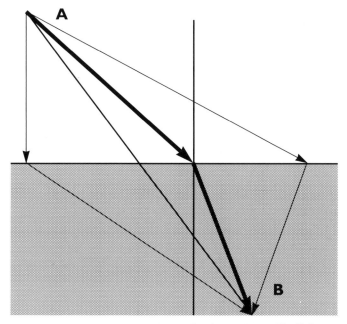

Figure 3 Illustrating Fermat's Principle of Least Time. The light ray travelling from A to B might travel along a number of different routes, as illustrated in the diagram. If the speeds of light in the two media are inversely proportional to the refractive indices *n*, it is easy to show that the path that takes the least time is that which satisfies Snell's law.

when account is taken of the different speeds of light in the two media, the shortest path is exactly that which satisfies Snell's law.

As we will see in a moment, this is only one way of deriving the law of refraction, but the key point is that the path of the light ray was derived by Fermat using what we now call a *Principle of Least Action*. It turns out that this is one of the most illuminating ways of formulating the laws of classical and modern physics – Newtonian physics, Lagrangian mechanics, quantum mechanics, quantum electrodynamics and the multiple universe theory of cosmology can all be formulated using different versions of the idea of considering all possible paths between A and B. The most likely path can be found from a Principle of Least Action and these have their roots in Fermat's profound insight.

Newton and the 'experimentum crucis'

Newton took his BA degree at Cambridge in 1665 and at about the same time the effects of the Great Plague began to spread north from London. The university was closed and Newton returned to his home at Woolsthorpe near Grantham where he entered what must be one of the most remarkably creative periods of anyone who has ever lived. As I tell my students, he was 22 at the time. In his own words, written over 50 years later:

> In the beginning of 1665, I found the method of approximating series and the rule for reducing the dignity [*power*] of any binomial into such a series. The same year in May, I found the method of tangents of Gregory and Slusius and in November had the direct method of fluxions and the next year in January had the theory of colours and in May following I had entrance to the inverse method of fluxions. And in the same year, I began to think of gravity extending to the orbit of the Moon and [...] from Kepler's rule of the periodic times of the planets being in sesquialternate proportion to their distances from the centres of their Orbs, I deduced that the forces which keep the planets in their Orbs must [be] reciprocally as the squares of their distances from the centres about which they revolve: and thereby compared the force requisite to keep the Moon in her orb with the force of gravity at the surface of the Earth, and found them [to] answer pretty nearly.
>
> All this was in the two plague years 1665–6. For in those days I was in the prime of my age for invention and minded mathematics and philosophy more than at any time since.
>
> (Additional Manuscript, Cambridge University Library)

In other words, in a matter of two years, he had discovered the binomial theorem and the differential and integral calculus in mathematics, the theory of colour in optics and the unification of celestial mechanics and the theory of gravity in astronomy. Even though it was a number of years before many of these reached their final definitive forms, the basic concepts were already formulated when he was only 22. It is a quite phenomenal achievement.

The theory of colour is our concern and, in particular, what came to be called the *experimentum crucis*. The history of this experiment is long and tangled and has recently been beautifully summarised by Simon Schaffer in his article 'Glass works – Newton's prisms and the uses of experiment'.

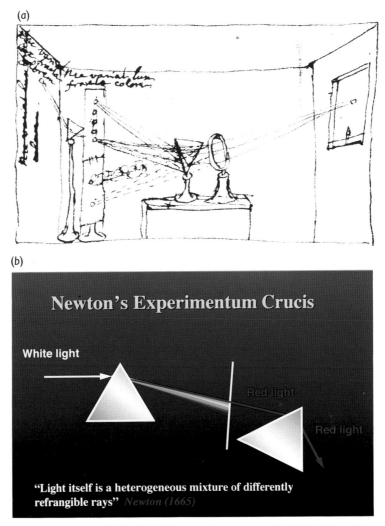

Figure 4 (*a*) Newton's sketch of his experimentum crucis. (*b*) A schematic diagram showing the decomposition of white light into the colours of the spectrum. On passing through the second prism, the colours are not split up into further colours.

Newton's first experiments were carried out as early as 1665. His sketch of the experiment is shown in Figure 4(*a*). It can be understood from this figure why commentators have remarked on Newton's genius as a scientist but his lack of facility as a draftsman. The experimental arrangement is shown in Figure 4(*b*). It was well known that light could be split into the colours of the rainbow by passing it through a prism, but, according to earlier investigators, it was believed that the process of passing the light through the prism modified it in some way. In the experimentum crucis, Newton selected a particular colour and passed it through a second prism. He found that, on passing that colour through the second prism, no further splitting of the light took place and so he concluded that light was split up into 'uncompounded colours' by the prism. Furthermore, he was able to recombine the dispersed light back into white light by reversing the prisms. In his own words, 'Light itself is a Heterogeneous mixture of differently refrangible rays'. In other words, white light is the result of adding together all the colours of the spectrum.

Newton was probably responsible for the concept that there are seven primary colours in the spectrum – he had a strong interest in musical harmonies and, since there are seven distinct notes in the musical scale, he divided up the spectrum into spectral bands with widths corresponding to the ratios of the small whole numbers found in the just scale. This idea was elaborated by Voltaire in his text *Élémens de la Philosophie de Neuton*.

The result of the experimentum crucis was hotly disputed both in England by Hooke and abroad by Huygens and the English Jesuits at Liège, Anthony Lucas and Edme Mariotte. The problem was that French scientists could not reproduce Newton's results. They claimed that the supposed primary colours could be further split into other colours. In a series of experiments, Mariotte claimed that a purely violet ray displayed red and yellow tinges after the second refraction. The reasons for the discrepancies were multifold, but partly, it was because it was difficult to obtain high quality prisms. In some of the experiments, the prisms had curved surfaces and in others they were of poor quality with bubbles and defects in the glass.

The upshot was one of the many bitter controversies in which Newton become embroiled throughout his life. Part of the problem was that Newton had not explained in any detail how he had conducted his experiments and he described only the successful experiments and not his unsuccessful attempts. It was recognised, however, that Newton was a gifted experimenter, as is testified by his design and construction of what we now call the Newtonian telescope. The optical design of this all-reflecting telescope was inspired by the problem that light is inevitably split into its constituent colours when it undergoes refraction through a refracting telescope. Newton had demonstrated that the different colours of the spectrum are refracted by different amounts and so it is impossible to bring white light to a single focus when large magnifications are involved. Not only did Newton design the all-reflecting telescope but he also ground the mirrors and constructed the telescope himself. It was shown, to great admiration, at the Royal Society and the improved power of the telescope was illustrated in the *Philosophical Transactions of the Royal Society* in 1672.

By the early 1700s Newton was President of the Royal Society and in a position of enormous power and patronage in scientific circles. His protégés initiated the cult of Newtonianism and one of these, Desaguliers, provided a critique of the French experiments. In 1715, a French delegation came to England to observe Newton's experimentum crucis. The carefully prepared experiments produced the desired result and were later successfully repeated in France. Desaguliers attributed the lack of success in the French experiments to their use of inferior prisms and claimed that the superior English prisms had to be used to observe the effect. The matter seemed conclusive but some scientists remained sceptical. As late as 1741, Rizzetti writes:

> It would be a pretty situation that in places where experiment is in favour of the law, the prisms for doing it work well, yet in places where it is not in favour, the prisms for doing it work badly.

This has a distinctly modern ring to it and it certainly is not a unique occurrence in the history of physics. The correctness of Newton's view enhanced his supreme authority in science and Newtonianism and Newtonian views on light and colour reigned supreme throughout the eighteenth century.

Atmospheric phenomena

The success of the experimentum crucis immediately led to the understanding of many natural optical phenomena. The simplest example is the origin of the colours of the rainbow. The prism experiments show that blue light is refracted more than red light and so the colours of the spectrum are displaced by different amounts on passing through spherical water droplets. It is a simple matter to produce an artificial rainbow using a spherical flask of water as a very large raindrop. Either the rays of the sun can be shone onto the flask and the rainbow projected onto the white card, or else, as illustrated in Figure 5, an intense light source can be used with a lens to produce a parallel beam of light which can be shone onto the flask. This simple experiment reveals all the properties of rainbows, including the increased light intensity within the arc of the rainbow. There are many intriguing variations on this theme. For example, light can be reflected more than once inside the spherical raindrops and this is the origin of the double rainbow (Figure 6(a)). Much more is going on in Figure 6(a), however, than a simple double rainbow. In this case the sun was setting over an expanse of water and so there is a second rainbow associated with the sunlight reflected from the water. The primary rainbows are associated with the deflection of light through an angle of 42°, as illustrated in Figure 6(b). In fact, each of these rainbows is double, the secondary rainbows being associated with the deflection of light through 51°, as illustrated in Figure 2(b).

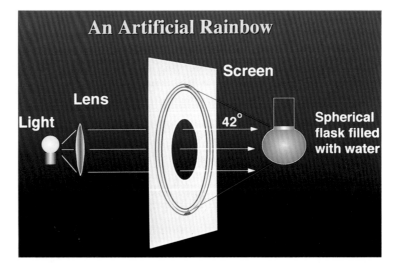

Figure 5 Illustrating how to create an artificial rainbow using a spherical flask of water. To produce the correct effect, the beam of light should consist of parallel light and should illuminate the flask uniformly.

This is just the beginning of a fascinating story of the many other types of atmospheric phenomena associated not just with droplets but with ice crystals. These phenomena are found particularly at northern and southern latitudes and, in these cases, the phenomena are associated

(a)

(b)

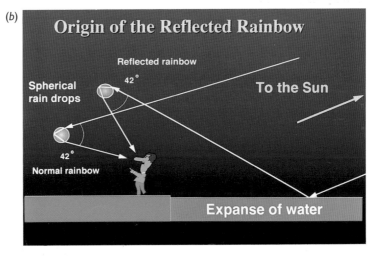

Figure 6 (a) An illustration of multiple rainbows produced when the sun is rising or setting over an expanse of water. The normal double rainbow can be observed with the colours running in opposite orders in the primary and secondary bows. In addition, a second double rainbow is observed that is due to the light reflected from the surface of the expanse of water. The origin of this phenomenon is illustrated in (b).

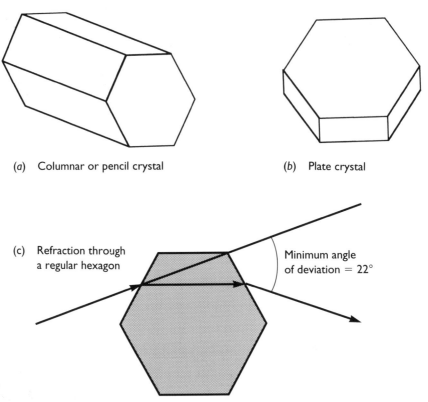

(a) Columnar or pencil crystal

(b) Plate crystal

(c) Refraction through
a regular hexagon

Minimum angle
of deviation = 22°

Figure 7 Illustrating the two forms of ice crystal: (*a*) the columnar or pencil crystal and (*b*) the plate crystal. (*c*) Illustrating the minimum angle of deviation of 22° through a regular hexagon.

with the refraction of light through ice crystals formed high in the atmosphere. The ice crystals are typically hexagonal, either in the form of plate crystals or what are known as columnar or pencil crystals (Figure 7(*a*) and (*b*)). Exactly the same type of analysis for light rays passing through hexagonal ice crystals can be performed as for spherical water droplets and again there is a minimum angle of deviation of the light through the hexagons. Ray tracing shows that light is concentrated towards the minimum angle of 22° (Figure 7(*c*)). This is the origin of the white circular halo sometimes seen about the sun at an angle of 22°.

To produce the most spectacular phenomena, it is necessary to align

Figure 8 An example of a remarkable display due to refraction through hexagonal ice crystals. The prominent bright images at the same latitude as the sun are known as sun-dogs. The details of this display can be accounted for by ray tracing through particular orientations of the ice crystals in the atmosphere.

the plate crystals quite precisely and this may occur naturally as plate crystals float down through the atmosphere. These aligned plate crystals are believed to be responsible for the formation of 'sun-dogs', the bright sun-like images seen at 22° from the sun but at the same altitude as the sun (Figure 8). If the pencil crystals are also lined up, some quite spectacular displays can be achieved. Even the very extreme example reported by Hevelius in Gdansk in 1661 can be accounted for by a suitable choice of the distribution of ice crystals. We can only wonder what early societies must have made of these incredible displays.

The theory of light

Newton did not have a particularly firm view about the nature of light. He favoured a corpuscular picture in which light consists of particles that travel from the object to the eye. In one version of the theory, the particles of light were supposed to be of the same size and the different coloured particles had different speeds. In another version, the different colours were considered to be associated with particles of different masses. These ideas reduced the processes of the reflection and refraction of light into collision problems in Newtonian dynamics. The theory was not developed to the high degree of mathematical exactitude that characterised his monumental work in mechanics and dynamics which was published in his *Principia Mathematica* of 1687.

A completely different point of view was being developed by Christian Huygens and published in his *Treatise on Light* in 1690. He considered light to be a form of wave which travels from the source to the observer. His great insight was to realise that each point in a wavefront of a wave can be considered to be the source of new waves with the same frequency of oscillation. It can be seen from the construction illustrated in Figure 9 that, if there are no obstructions, the wavelets associated with a particular wavefront add up coherently only in the forward direction.

The great achievement of Huygens' theory was the explanation of the phenomena of reflection and refraction of light. Unlike Descartes, he believed light travelled more slowly in material media than in air and so the wavelength of the waves of a fixed frequency was shorter and the wavelets added up at a different angle, accounting for the phenomenon of refraction. In fact, Huygens' construction is still by far the simplest and most useful way of understanding many of the properties of light, as we will see in a moment.

Battle was immediately joined with Newton. The conflict between the wave and particle theories of light became entangled with Newton's vitriolic feud with Leibniz over the discovery of the integral and differential

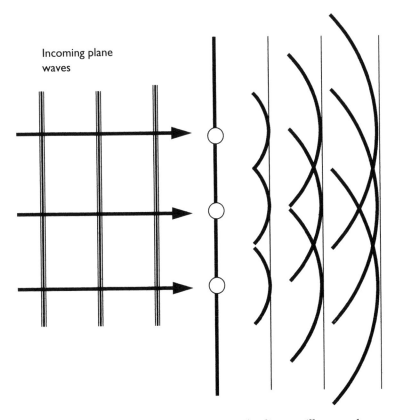

Incoming plane
waves

Figure 9 Illustrating Huygen's construction. The diagram illustrates how a plane wave can be represented by the superposition of wavelets originating in each wavefront. The diagram shows only three points on the wavefront that are considered to be the sources of wavelets. These wavelets add up coherently only in the direction of propagation of the wave. If the number of points were increased to infinity, the resulting wavefront would be planar.

calculus. Throughout the eighteenth century, victory went to Newton, not so much through his own work but rather through the enthusiasm of his acolytes – what have been called Newton's stormtroopers. The power of Newtonian mechanics and dynamics was universally recognised and Newtonianism spread throughout Europe. I have already remarked on Voltaire's text on Newton's natural philosophy. Another lovely example is Francesco Algarotti's text of 1743 *Newtonianismo per le dame*, providing a detailed account of Newton's theory of light for ladies.

The triumph of the wave theory

Huygens' views were far ahead of their time. Only in the first half of the nineteenth century did the pendulum began to swing in favour of the wave theory. Among the pioneers was Thomas Young, who made two central contributions to the subject of this chapter. He undertook some of the first experiments on the interference of light and, in addition, he proposed that any colour could be obtained by mixing together lights of only three different colours: the three-colour theory of colour vision.

His most famous experiment is the double-slit experiment, in which light is passed through two small holes or slits which are very close together. If we project the light onto a screen close to the holes, we observe two spots of light but, as the screen is moved further and further away, we begin to observe the interference of one beam of light with the other and, rather than the image of the two slits, a pattern of light and dark fringes is observed (Figure 10(a)). This can be readily explained by the wave theory of light using Huygens' construction. If a plane wave is incident upon the holes, we can replace the holes by two sources of light and then we can work out simply the interference of the waves from these two sources. It can be seen in Figure 10(b) that the waves add up constructively along certain directions and cancel out along others, the result being the form of interference pattern illustrated in Figure 10(a).

The case for the wave theory became overwhelming during subsequent years. Among the most important experiments were those of the French physicists to measure precisely the speed of light. The first accurate value was measured by Fizeau in 1849. In 1850, this was immediately superceded by Foucault's brilliant experiments, originally proposed by Arago, in which he showed that the speed of light is less in water than in air by exactly the ratio of their refractive indices. The observation of the polarisation of light was crucial in leading to the concept that light consists of *transverse waves*. When waves travel along a taut string, the displacement of the string is sideways; that is, in a direction transverse to the direction of propagation of the wave along the string. The direction of the displacement defines what is knows as the *polarisation* of the wave. Suppose we place a very narrow slit in the path of the wave propagating down the string. Then, if the string vibrates in a direction parallel to the

(a)

Screen

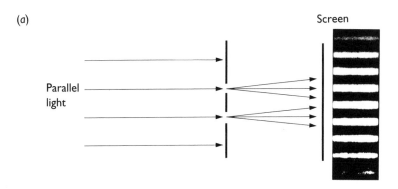

Parallel light

(b)

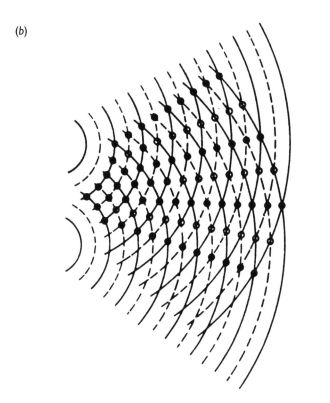

Figure 10 (*a*) and (*b*) Illustrating the interference of light waves in Young's double slit experiment. The waves add up constructively along particular angles and destructively along intermediate angles. This results in the typical interference pattern illustrated at the upper right.

slit, the wave can pass through the slit, but, if the vibration is perpendicular to it, the wave cannot pass through but is reflected. Light behaves in a similar manner – Polaroid® used in sunglasses is a simple example of a material which allows one of the polarisations of light to pass through to the eye. Thus, light was considered to be some form of transverse wave.

But there is a problem – what is the medium through which the waves are travelling? We can observe the light from the distant stars and there is very little matter indeed in the space between them and us and yet light has no problem in travelling at the same constant speed between the star and the Earth. The wave theories gave rise to the concept of the *Aether* as the invisible medium through which light waves propagate. The search for the aether is one of the most tantalising stories of nineteenth century physics, and the resolution of that problem led to completely new concepts of the nature of space and time.

Maxwell and the theory of colours

James Clerk Maxwell is Scotland's greatest physicist and, in my view, his contributions to the whole of physics rank almost with those of Newton. Here we can only touch on his contributions to the understanding of the nature of light and colour. Like Newton and Young before him, he made central contributions to the understanding of both subjects.

(a)

Figure 11 (a) A photograph of Maxwell's portable light box.

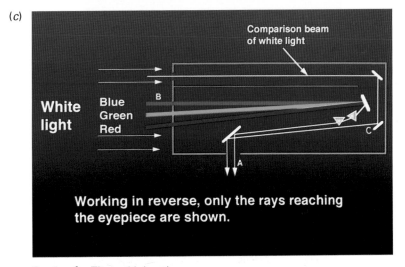

Caption for Figure 11 *(cont.)*
(*b*) Illustrating the principle behind the light box. If white light is passed through the prisms from A to B, the spectrum is displayed as shown. (c) If white light is now incident upon the slits at B, only the indicated colours fully illuminate the prism when observed from A. By varying the width of the slits, the amounts of red, green and blue light can be varied to produce different colours at A. For C, see the text.

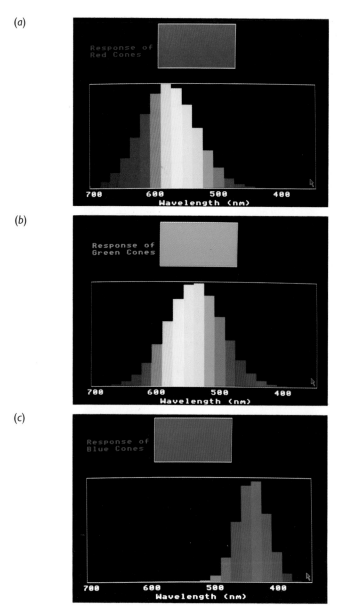

Figure 12 The sensitivity of the (*a*) red, (*b*) green and (*c*) blue cones in the retina of the eye to different wavelengths of light.

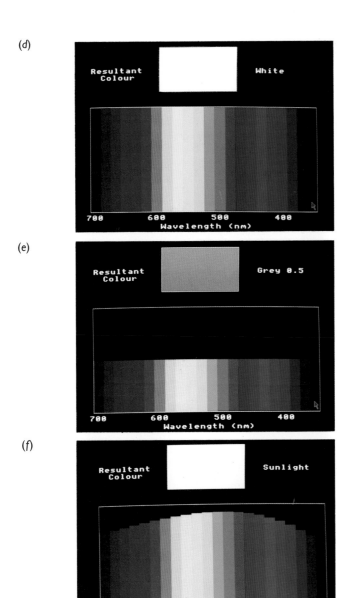

Caption for Figure 12 (*cont.*)
The response of the eye to different spectral distributions of light: (*d*) white light is the equal combination of all the spectral colours; (*e*) grey colours are produced by reducing the intensity of white light; (*f*) the spectrum of the sun is remarkably close to a white light spectrum.

(g)

(h)

(i)

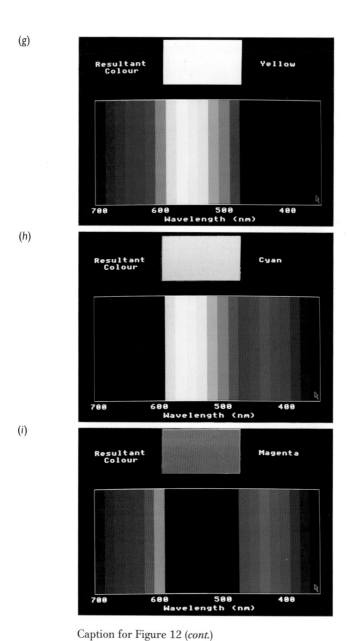

Caption for Figure 12 (*cont.*)
(*g*) by eliminating short wavelengths yellow light is observed; (*h*) by eliminating long wavelengths cyan is observed; (*i*) by eliminating the middle wavelengths magenta is observed.

(j)

(k)

(l)

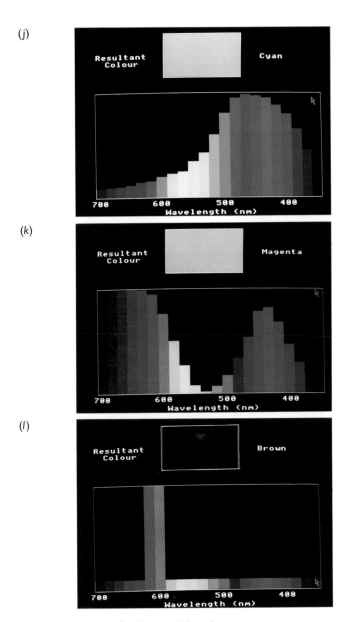

Caption for Figure 12 (*cont.*)

Actual pigments do not have the ideal absorption spectra: for example, a cyan pigment might have the absorption spectrum illustrated in (*j*), or magenta the spectrum shown in (*k*). (*l*) Brown is produced by adding orange to grey.

Maxwell had been fascinated by colours since his childhood. In an early photograph, he can be seen with his famous colour wheel, which he used to show how white light could be produced by different combinations of colours. His major contributions to the theory of colour were inspired by what had become known as three-colour theory of Thomas Young and Hermann von Helmholtz. This theory held that light of any colour could be produced by the appropriate mixing of only three different colours, which span the range of the visible spectrum. Conventionally these can be thought of as the red, green and blue regions of the spectrum. It was well into the twentieth century before the physiological basis for this concept was fully understood (see Baylor, Chapter 4, and Mollon, Chapter 5). In the eye there are two types of light receptor called *rods* and *cones*. The cones come in three different types which are sensitive to red,

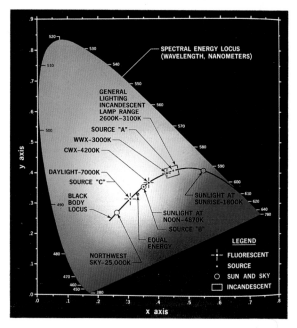

Figure 13 The standard CIE diagram for describing quantitatively the composition of different colours. The pure spectral colours are shown round the upper perimeter of the diagram. Along the bottom edge are colours produced by mixing pure red and blue. If three colours are selected, other colours can be synthesised by adding amounts of the colour inversely proportional to the distance of the point from the three selected colours.

green and blue light. The eye can distinguish quite incredibly small differences in the relative intensities of the light detected by the three types of colour receptor, and it is these tiny differences that give rise to the huge range of different colours which the brain can recognise.

In the late 1850s Maxwell worked out quantitatively how to synthesise different colours using only red, green and blue light using a piece of apparatus which was known as *Maxwell's light box*. Maxwell's original portable light box was built by Messrs Smith and Ramage of Aberdeen in 1860 (Figure 11(*a*)). The principle of operation of the light box is shown in Figure 11(*b*) and (*c*). If white light enters the box through the hole marked A, it is reflected onto the double prism, where it is split up into its spectral components and then reflected down the box to the far end, where there are three adjustable slits B to allow different pure colours to leave the box. In a typical experimental arrangement, red, green and blue light would be selected, as shown in Figure 11(*b*). Now Maxwell realised that if instead he shone parallel white light on to the slits at B, the only colours which reach A are red light from the 'red' slit, green light from the 'green' slit and blue light from the 'blue' slit (Figure 11(*c*)). Furthermore, each of these colours fully illuminates the prism as observed from A. The beauty of the experiment is that, by adjusting the widths of the slits at B, he could superimpose precisely known amounts of red, green and blue light as oberved at A. In addition, he allowed undispersed white light to arrive at A by the route C to act as a comparison light source. In this way, Maxwell developed the quantitative theory of three-colour vision.

I have written a simple computer program to illustrate how the theory of colours works quantitatively. The visible spectrum extends from wavelengths of about 700 nanometres (nm) at the extreme red end of the spectrum to about 400 nm at the blue end. The sensitivity curves of the red, green and blue cones of the eye are shown in Figure 12(*a*), (*b*) and (*c*). We have synthesised the colours of the spectrum by adding together red, green and blue lights as they would be observed by the eye. For the purposes of this illustration, we have somewhat enhanced the far red and far blue regions of the spectrum. What the program does is to show the colour we observe with our eyes when we change the incident spectrum of radiation. First of all, if we have an incident spectrum which has the same intensity at all wavelengths, we observe white light (Figure 12(*d*)). If we reduce the intensity of the light equally at all wavelengths, we observe deeper and deeper greys, as illustrated in Figure 12(*e*). Now, the light emitted by the sun is remarkably similar to this flat intensity spectrum. It can be represented by a thermal or black-body spectrum at a temperature of 5700 kelvin (K) and this is shown in Figure 12(*f*). It can be seen that it is more or less indistinguishable from a pure white light spectrum.

We can now simulate the effects of mixing colours or lights to produce different perceived colours. If we cut out the short wavelength region of the spectrum, from 400 to 500 nm, we obtain a yellow colour (Figure 12(*g*)); we can think of this colour as the sum the separate colours shown in the figure or else as the reflected spectrum of white light from a pigment that absorbs perfectly in the blue region of the spectrum. In the same way, if we cut out the long wavelength range, 600 to 700 nm, we obtain the colour cyan (Figure 12(*h*)); and if we cut out the central one third of the spectrum, 500 to 600 nm, we obtain magenta (Figure 12(*i*)). These three examples illustrate how pigments or paints modify the light we observe in a painting. Pigments absorb certain ranges of wavelength and reflect others; for example, magenta has the property of absorbing the central wavelengths of the visible spectrum and reflecting back to the viewer the other colours. Thus, whereas with light, we create different colours by the process of the *addition* of different lights, with pigments we observe different colours through the process of the *subtraction* of different colours from the incident spectrum. Now, in practice, real pigments do not have the perfect absorption and reflection properties illustrated in Figure 12(*g*), (*h*) and (*i*). The properties of real pigments are more like those illustrated in Figure 12(*j*) and (*k*). The important point of principle is that the eye has only three different colour sensors and so, if each type of cone absorbs the same intensity, no matter how that light was distributed through the incident spectrum, we observe the same colour.

The above discussion has been concerned principally with the mixing of lights to produce different colours or *hues*. In addition, the eye is sensitive to the *brightness* of the colour, in other words how intense the colour is, as was illustrated by the example of reducing the intensity of the light from white to grey (Figure 12(*d*) and (*e*)). Finally the eye is also sensitive to its *saturation*; that is, whether it is a pure colour in the sense of Newton (saturated) or whether it is observed against a grey background (unsaturated). The last example Figure 12(*l*) shows how brown can be synthesised by adding orange to a grey background.

This discussion leads us to the modern quantitative theory in which the colours of the spectrum are represented on what is known as a CIE (Commission Internationale d'Éclairage) diagram. In this diagram (Figure 13), the pure colours of the spectrum are represented around the perimeter of the coloured region of the diagram running from extreme blue (380 nm) at the bottom left round the top of the diagram through green at 520 nm to the extreme red end of the spectrum (780 nm) at the bottom right. The colours represented along the bottom of the diagram are not pure colours but are obtained by mixing different amounts of red and blue light. The advantage of the CIE diagram is that we can obtain quantitative information about how much light of different pure colours is needed to synthesise any of the colours represented on the diagram. The details of the use of this diagram would take us on a major digression and I can recommend the book *Light and Colour in Nature and Art* by Williamson and Cummins where these and many other issues are discussed. For example, we have given no consideration to the nature of the surfaces from which the light is reflected – are they smooth, rough, highly reflective, polarising and so on?

From the point of view of physics, there is not a great deal to add to this story. However, this is where physiology and the theory of colour perception begins, and these are dealt with in Chapter 4 by Denis Baylor and Chapter 5 by John Mollon.

Maxwell and the theory of the electromagnetic field

We now turn to Maxwell's theory of the electromagnetic field. This is not the place to go into the details of his brilliant theoretical analysis of electromagnetism, but let me describe simply the process of visualisation which led to his initial discovery of the theory of the electromagnetic field.

The laws of electromagnetism had been built up gradually through the first half of the nineteenth century. The laws of electrostatics and magnetostatics had been established through the experimental work of

Coulomb, and the mathematical apparatus to describe these phenomena
was developed by Poisson. The production of magnetic fields by currents
had been demonstrated by Oersted and quantified by Ampère, Biot and
Savart. In 1831, Faraday made his momentous discovery of the law of
electromagnetic induction and this was put into mathematical form by
Weber. The theory of the electromagnetic field was, however, incomplete.

Maxwell turned to this problem in the early 1860s. From his youth,
Maxwell was impressed by the power of analogy as a tool for uncovering
the mathematical laws of physics, and he now used it in a most extreme
form. By analogy with fluid flow, he imagined the magnetic field to consist of a system of rotating vortices in the aether. His model for *material
media* or a *vacuum* containing a magnetic field is shown in Figure 14. The
vortices are represented by regular hexagons in this sketch, which
Maxwell published in the *Philosophical Transactions of the Royal Society*.
The strength of the magnetic field was assumed to be proportional to the
speed of rotation of the vortices. If left on their own, however, the vortices
will be disrupted and dissipated as a result of friction and so Maxwell
inserted 'ball-bearings' between them to act as idler wheels.

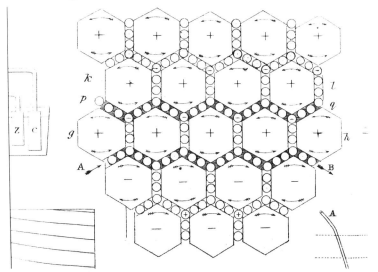

Figure 14 Maxwell's model of the aether, from which he derived the
concept of the displacement current. Using this model, he showed that the
speed of propagation of electromagnetic disturbances in a vacuum is the
speed of light.

He then identified the ball-bearings with electric particles, which, if they were free to move, would carry a current when an electric field was applied. In insulators or, more significantly, in a vacuum, the electric particles are not free to move but they can be displaced from their stationary positions by the action of an electric field. This was the mechanical model that Maxwell used to work out his equations for the electromagnetic field in their primitive form. The key step was the introduction of the concept that the electric particles are *displaced* from their normal positions by the action of an electric field. When they move under the action of the field, they give rise to what he called a *displacement current.*

Now, Maxwell took the crucial step. He realised that waves would be able to propagate through this system of vortices and ball-bearings – if the electric particles are displaced, they cause a disturbance that propagates through the medium, and he was able to work out the speed at which these waves travel. To his astonishment, he found that the speed of the waves was entirely determined in terms of the fundamental constants of electrostatics and magnetostatics and furthermore that it was exactly the speed of light.

This is one of the greatest discoveries of theoretical physics: electromagnetic disturbances propagate at the speed of light through a vacuum, and Maxwell did not hesitate to identify light as electromagnetic waves. He had discovered the nature of the waves which lie behind the wave theory of light. Maxwell became as excited about this result as he ever became about anything. In a letter to his cousin Charles Cay, in what Everitt calls 'a rare moment of unveiled exuberance', he wrote:

> I have also a paper afloat, containing an electromagnetic theory of light, which, till I am convinced to the contrary, I hold to be great guns.

In the following year, the theory of electromagnetism was completely rewritten independent of the creaky scaffolding by which it had been erected and Maxwell's equations of the electromagnetic field were revealed in all their glory for the first time.

I like to tell students about this story since it illustrates how the really great scientists have the ability to use the most remarkable methods and techniques to gain new insights. The approach certainly does not appeal to everyone. Poincaré was heard to remark that in his experience all

Frenchmen were oppressed by a 'feeling of discomfort, even of distress' at their first encounter with the works of Maxwell.

Maxwell's was a staggering achievement but it brought with it as many problems as it solved. There were two in particular which should be noted. Maxwell had discovered electromagnetic field equations which had solutions for waves propagating *in a vacuum*. What has happened to the aether if waves are able to propagate through completely empty space? Furthermore, if there is nothing but a vacuum, with respect to what are the waves supposed to travel at the speed of light?

The second point is more subtle. According to Newtonian mechanics, the equations which Maxwell derived turned out to be valid in only one special frame of reference. In other words, if the laws of electromagnetism were correct in one frame of reference, say the laboratory frame of reference, they would no longer have the correct form if the same experiment were performed in, say, a laboratory placed on a uniformly moving train. For this reason, Maxwell's equations were originally thought to be 'non-relativistic'. Maxwell died in 1879 and did not live to see the complete validation of his theory of the electromagnetic field by the experiments of Hertz.

Heinrich Hertz's experiments

The validation of Maxwell's theory came in the late 1880s when Heinrich Hertz began his great series of experiments on the propagation of electromagnetic phenomena in air; in other words, direct tests of Maxwell's prediction that electromagnetic phenomena should behave exactly like light. The problem was to construct emitters and receivers of electromagnetic disturbances. He used electrostatic spark devices to produce short-wavelength radio waves and, in his final experimental arrangement, he produced waves with wavelengths as short as 30 cm. His emitter and receiver are shown in Figure 15(A).

The emitter (a) produces electromagnetic disturbances by electric discharges between the spherical conductors. The detector (b) consists of a similar device with the jaws of the detector placed as close together as possible so as to achieve maximum sensitivity. The emitter was placed at the focus of the cylindrical paraboloid reflector to produce a directed

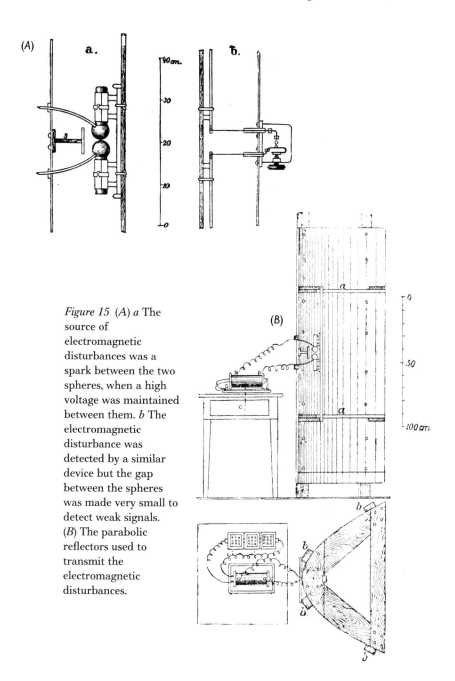

Figure 15 (A) *a* The source of electromagnetic disturbances was a spark between the two spheres, when a high voltage was maintained between them. *b* The electromagnetic disturbance was detected by a similar device but the gap between the spheres was made very small to detect weak signals. (B) The parabolic reflectors used to transmit the electromagnetic disturbances.

beam of radiation (*B*). With this apparatus, Hertz demonstrated that electromagnetic waves had all the properties of light – linear propagation, reflection, refraction, polarisation. The speed of propagation of the waves was measured by the product of their wavelength and their frequency and this turned out to be exactly the speed of light. This work was completed by 1889 and was the ultimate triumph for the wave theory of light and for Maxwell's theory of the electromagnetic field.

The whole structure of physics was, however, just about to crumble. In the very same series of experiments, Hertz also discovered in 1887 a key phenomenon that was to lead to the downfall of classical physics. In his experiments on spark discharges from metal electrodes, he discovered that when ultraviolet light shone on the electrodes, the discharge took place at a lower voltage. This is the phenomenon known as the *photoelectric effect* and it was soon shown that the cause of this was electrons ejected from the metal of the electrodes by the incident ultraviolet radiation. It turned out to be very difficult to understand the photoelectric effect in terms of classical physics. What was particularly strange was that, as the intensity of light incident upon the electrodes was increased, the energies of the ejected electrons did not change – all that happened was that more electrons were emitted with the same energies.

Planck and Einstein

There was, however, much more going wrong with classical physics at this time. To list but a few of the problems:

1. Classical physics was unable to give an account of the complete spectrum of the thermal radiation from a hot body – what is known as the spectrum of black-body radiation.
2. It could give no satisfactory explanation of the emission and absorption lines that appear in the spectra of atoms and molecules.
3. It gave the wrong answers for the heat capacities of gases.
4. The Michelson–Morley experiment revealed no evidence for the aether.
5. The photoelectric effect could find no adequate explanation in terms of classical physics.

For me, the story of the resolution of these problems is one of the most dramatic and heroic stories in the whole of physics and indeed of modern culture. Within a matter of years the whole foundation of physics was undermined. Once we understand the enormous intellectual struggles involved, we obtain a better understanding of what science is really about than through anything else I know of. It is no exaggeration to say that the resolution of these problems through the discoveries of relativity and quantum mechanics are the basis of much of twentieth century civilisation as we know it. It is a story which deserves to be much better known.

To cut a long story short, in 1900 Planck introduced the concept of *quantisation* in order to account for the spectrum of black-body radiation. He found that the only way in which he could account for the spectrum was to assume that the emitters of the radiation did so only between well-defined energy states rather than between all possible states.

Einstein went very much further. In 1905, he published three of the greatest papers of modern physics. At that time he was 26 and employed as a technical expert, third class, at the Swiss patent office in Bern. Any one of these papers would have ensured him a position in the pantheon of great physicists. The titles of these three papers are as follows:

1. On the theory of Brownian motion.
2. On the electrodynamics of moving bodies.
3. On a heuristic viewpoint concerning the production and transformation of light.

The first demonstrated conclusively the molecular nature of matter. The second is the famous paper on the special theory of relativity. Light plays the central role in the new understanding of the nature of space and time. The essence of that revolution is encapsulated in the statement that *the speed of light is a constant in free space under all circumstances.* This immediately makes the question of the nature and existence of the aether completely redundant. The speed of light assumes the role of an absolute limiting speed. Exactly this same postulate leads to the four-dimensional view of the world embodied in what we now call *space–time* and the famous relation between mass and energy, $E = mc^2$.

This is remarkable enough but it was nothing compared to what came in the third paper. In Einstein's own words, the paper is 'very revolutionary', words which he did not use about his paper on special relativity. Let me quote the opening paragraphs:

There is a profound formal difference between the theoretical ideas which physicists have formed concerning gases and other ponderable bodies and Maxwell's theory of electromagnetic processes in so-called empty space. Thus, while we consider the state of a body to be completely defined by the positions and velocities of a very large but finite number of atoms and electrons, we use continuous three-dimensional functions to determine the electromagnetic state existing within some region, so that a finite number of dimensions is not sufficient to determine the electromagnetic state within that region completely . . .

The undulatory theory of light, which operates with continuous three-dimensional functions, applies extremely well to the explanation of purely optical phenomena and will probably never be replaced by any other theory. However, it should be kept in mind that optical observations refer to values averaged over time and not to instantaneous values. Despite the complete experimental verification of the theory of diffraction, reflection, refraction, dispersion and so on, it is conceivable that a theory of light operating with continuous three-dimensional functions will lead to conflicts with experiment if it is applied to the phenomena of light generation and conversion.

Einstein proposed that, for some purposes, it would be more appropriate to consider light to consist of particles rather than waves. This was pretty inflammatory stuff. Hertz had just confirmed Maxwell's theory of the electromagnetic field and now Einstein was proposing to replace all of that magnificent achievement with the idea that light actually consists of particles. Even Planck would not go that far and was initially opposed to Einstein's proposal. But, gradually, the idea gained ground. Specifically, Einstein proposed to explain the photoelectric effect as a collision between particles of light and the electrons of the material, and he made a very strong prediction about the results of precise measurements of the energies of electrons ejected by the particles of light of different frequencies in the photoelectric effect. In 1916, Millikan showed that Einstein's prediction was precisely correct (Figure 16). In Millikan's words:

> We are confronted however by the astonishing situation that these facts were correctly and exactly predicted nine years ago by a form of quantum theory which has now been generally abandoned.

He refers to Einstein's 'bold, not to say reckless, hypothesis of an electromagnetic light corpuscle of energy $h\nu$ which flies in the face of the thoroughly established facts of interference.'

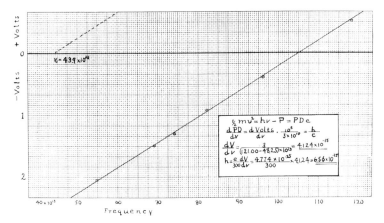

Figure 16 Millikan's measurements of the energies of photoelectrons liberated by different incident wavelengths of light. The straight line passing through the points shows Einstein's predictions, made roughly ten years earlier.

The particle nature of light can be demonstrated remarkably easily nowadays using a highly sensitive charge coupled device (CCD) camera in conjunction with the double-slit experiment. When Young's double-slit experiment (Figure 10) is repeated with high light intensities, we observe the usual appearance of interference fringes, but, as we decrease the light intensity, we begin to see the pattern of fringes wobble about. This is because the individual particles of light or photons are arriving at the screen and they arrive at random times. Where the photons arrive, however, is not random. When we add up where many particles arrive, we obtain the standard interference pattern found by Young. Thus, we can demonstrate directly that light can also be considered to consist of photons. But, if they are particles, how did they know to land in the correct positions to form a diffraction pattern rather than simply produce a shadow of the holes?

In the 1920s, two further great experiments were carried out. In 1923, Compton demonstrated the change in energy of a high energy X-ray photon when it collides with a stationary electron. In these collisions, the electron gains energy and the energy (or frequency) of the photon decreases. In 1927, Davisson and Germer showed that not only do electromagnetic waves have particle-like properties but electrons have wave-like properties too. This concept had been proposed by de Broglie as early as 1924 and it proved to be the inspiration behind Schrödinger's formulation of wave mechanics. Within a couple of years, the new foundations of physics as embodied in wave and quantum mechanics were established.

Conclusion

Where does this leave us? What is light? We can describe very precisely what is does:

1. It has wave-like properties that give very exact answers according to the classical theory of interference.
2. It has particle-like properties that give very exact answers in collision processes between photons and particles.
3. It travels through a vacuum at the constant speed of light.
4. It has the same constant speed in a vacuum no matter how we look at it, be it in a laboratory frame of reference or from a space ship moving close to the speed of light.
5. The particles of light have no mass.

None of this causes physicists to lose any sleep at all. We do not even ask why it should be so – *that is just how nature is*. To describe these properties quantitatively, we need the theories of relativity and quantum mechanics. These are combined in a quite superb theory known as *quantum electrodynamics*, which can describe very beautifully all of these properties and is the most precise theory that has been tested in the whole of physics. Richard Feynman has described how it all fits together in a delightful little book entitled *QED*, standing for Quantum Electrodynamics, in which he explains the essence of the theory in lay language.

I think of the case of light as being a paradigm for the whole process of modern physics. So long as one regards the facts of experiment and observation as the 'reality' of nature, the theoretical apparatus is a precise, predictive machinery for making imaginative application of the laws to the diverse ways in which Nature works. And remember, light would still have all these properties, even if we didn't have the mathematical apparatus to describe it.

Acknowledgements

During the lecture, I performed six experiments or demonstrations. I am most grateful to Professors John Baldwin and Richard Hills, and Messrs Alan Chapman, Charles Bird and Brian Clark for assistance in preparing the demonstrations. I am also most grateful to Mr Andrew Blain, who assisted me in performing the experiments during the lecture. My sincere thanks also go to Mark Longair, who did all the difficult bits of programming for the computer simulations of the three-colour theory of colour vision.

Further reading

Campbell, L., and Garnett, W., *The Life of James Clerk Maxwell*, p. 342 (Letter of 5 January 1865), London: Macmillan & Co., 1882.

Einstein, A. *Annalen der Physik*, **17** (1905), 132.

Fauvel, J., Flood, R., Shortland, M., and Wilson, R. (eds.), *Let Newton Be: A New Perpsective on his Life and Works*, Oxford: Oxford University Press, 1989.

> This contains a remarkable set of essays describing many of the lesser known aspects of Newton's life and career.

Feynman, R. P. *QED: The Strange Theory of Light and Matter*, Princeton, NJ: Princeton University Press, 1985.

Gooding, D., Pinch, T., and Schaffer, S. (eds.), *The Uses of Experiment: Studies in the Natural Sciences*, Cambridge: Cambridge University Press, 1989.

> As the editors remark, 'Experiment is a respected but neglected activity'. This book contains many essays on the role of experiment in the natural sciences. Simon Schaffer's splendid article on Newton's experimentum crucis was used extensively in the preparation of this chapter.

Greenler, R., *Rainbows, Halos and Glories*, Cambridge: Cambridge University Press, 1991.

> This volume contains beautiful illustrations of natural optical phenomena and presents computer simulations of some of the most complex displays.

Longair, M. S., *Theoretical Concepts in Physics*, Cambridge: Cambridge University Press, 1992.

> I include my own book in the reading list because it contains a chapter describing the origin of Maxwell's equations and five chapters on the work of Planck and Einstein, which led to the discovery of quanta. Although it contains details of the theoretical arguments, the essence of the story can be obtained from the narrative parts.

Millikan, R. A., 'A direct photoelectric determination of Planck's "h"', *Physical Review*, **7** (1916), 355–88.

Minnaert, M., *The Nature of Light and Colour in the Open Air*, New York: Dover Publications Inc., 1954.
    This is a classic text which explains in simple language optical phenomena outdoors. It also suggests many fascinating simple experiments and observations.
Rizzetti, G., *Saggio dell'antinewtionianismo sopra le Leggi del Moto e del Colori*, p. 112, Venice, 1741.
Turnbull, H. W., and Scott, J. F. (eds.), *Correspondence of Isaac Newton*, vol. I, p. 95, Cambridge: Cambridge University Press, 1957.
Westfall, R. M., *Never at Rest*, Cambridge: Cambridge University Press, 1987.
    This is the classic biography of Isaac Newton.
Williamson, S. J., and Cummins, H. Z., *Light and Colour in Nature and Art*, New York: John Wiley and Sons, 1983.
    This is a gentle introduction to the physics of a very wide range of topics in the general area of light and colour. It includes chapters on the physics of different methods of painting as well as details of modern colour systems.

4   Colour Mechanisms of the Eye

Denis Baylor

The rich colours that we see are inventions of the nervous system rather than properties of light itself. Colour, like beauty, is in the eye and brain of the beholder. Early in the last century Thomas Young, who elucidated the wave nature of light, recognised the all-important role of the observer in colour sensations. He proposed that colour was an internal phenomenon resulting from the activity of three resonators – tuning forks – each set into vibration by light in a particular part of the visible spectrum. In 1801 he said:

> As it is almost impossible to conceive each sensitive point of the retina to contain an infinite number of particles, each capable of vibrating in perfect unison with every possible undulation, it becomes necessary to suppose the number limited; for instance to the three principal colours, red, yellow and blue, and that each of the particles is capable of being put into motion more or less forcibly by undulations differing less or more from perfect unison. Each sensitive filament of the nerve may consist of three portions, one for each principal colour.
>
> (MacAdam, 1970)

My purpose in this chapter is to examine how the eye derives colour information and represents it in a form that can be transmitted to the brain. Young's resonators will take centre stage. After briefly considering light itself I will describe how it starts the visual process. Next I will present evidence that Young's resonators consist of three kinds of light-absorbing pigment molecule contained within the photoreceptor cells of the eye. After looking at the structure and genetics of the pigment molecules I will examine how their relative activities, which encode colour, are signalled to the brain.

Preliminaries

The light energy that initiates colour sensations has two fundamental dimensions: intensity and wavelength. The intensity determines how bright a light appears. Since light energy is transferred in discrete packets of energy – photons, or quanta (see also Longair, Chapter 3) – the intensity can be specified as the number of photons that fall on a given cross-sectional area per unit time. This is equivalent to expressing the intensity of rain by giving the number of drops that fall each second on a roof tile of given size. The wavelength of the light is the distance between successive crests in the electromagnetic wave train. Wavelength determines whether the light can be seen at all, as well as the colour sensation that it evokes.

Only light at wavelengths between roughly 400 and 700 nanometres (nm) can be seen because only these wavelengths are absorbed in the photoreceptor cells of the eye. In this band of wavelengths photons from the sun are most plentiful at the surface of the earth, suggesting that the eye's sensitivity has evolved to match the abundance of the photons available for seeing. As the wavelength increases from 400 nm, the perceived colour changes progressively from blue to green to yellow to red. It is remarkable that small quantitative changes in the physical parameter wavelength can produce striking qualitative changes in the perceived colour: light at 550 nm appears green, while light at only 5% longer wavelength (580 nm) appears yellow. The essence of colour vision is to sense wavelength and separate it from intensity.

Wonderful as it is, normal human colour vision has striking limitations that can be brought out by colour-matching experiments. Consider the following example (Figure 1). A subject views a screen onto which is projected a narrow band of wavelengths near 580 nm. This 'monochromatic' light appears yellow. An identical sensation of yellow can be produced by superimposing patches of monochromatic red light (620 nm) and green light (540 nm). Only when monochromatic lights are presented one at a time is there a unique relation between wavelength and perceived colour. How can we explain the equivalence of the monochromatic (580 nm) yellow and the mixture (540 and 620 nm)? Furthermore, in some individuals, genetic variations impose even stronger limitations. To some men, for

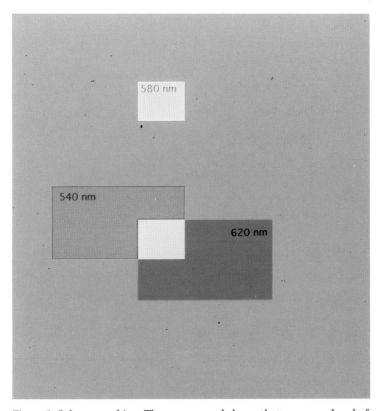

Figure 1 Colour matching. The upper panel shows that a narrow band of wavelengths near 580 nm appears yellow when projected onto a screen. The patches of light below, at 540 and 620 nm, appear green and red, respectively, but where they superimpose they appear yellow. With suitable adjustment of the light intensities, the yellow below can be made to match exactly that above.

example, a monochromatic yellow looks identical with *either* 540 nm light *or* 620 nm light. What has gone wrong here?

First signals of vision

Vision begins in the retina, the thin plate of nerve cells lining the inside of the eyeball. At the back of the retina is a layer of photoreceptor cells which absorb light and generate the neural signals that initiate vision. There are two types of photoreceptor, rods and cones (Figure 2). Rods are so exquisitely sensitive that they signal the absorption of a single photon.

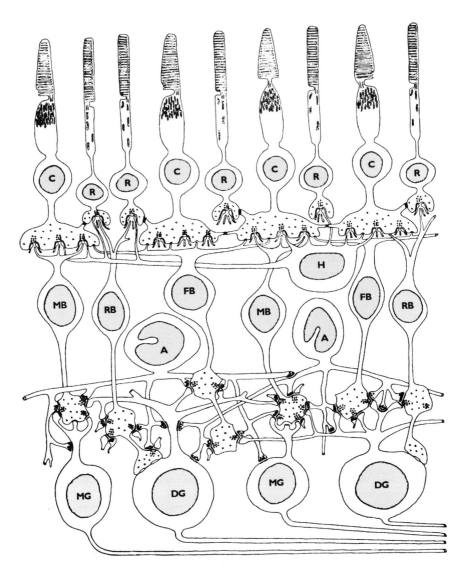

Figure 2 Diagrammatic cross-section of the retina, showing rod (R) and cone (C) photoreceptor cells, bipolar cells (RB, FB, and MB), horizontal cells (H) and amacrine cells (A). Fibres of retinal ganglion cells (MG, DG) form the optic nerve. Direction of the light path is upwards.

They mediate vision in very dim light such as starlight. Cones are less sensitive and mediate vision in ordinary daylight. Colour sensations are their unique province. The rods and cones make synaptic contacts on bipolar cells. These second-order neurons carry the visual message to the retinal ganglion cells, which lie at the inner surface of the retina. Long fibres from these cells carry messages to the brain over the optic nerve. Horizontal and amacrine cells regulate information flow from photoreceptors to bipolar cells, and from bipolar cells to ganglion cells, respectively.

Rods and cones have similar architecture. Light is absorbed in visual pigment molecules: rhodopsin in the rods, and one of three cone pigments in the cones. The pigment is located in membranes within the outer segment, a long cylindrical process at the end of the light path through the retina. An outer segment contains roughly 100 million pigment molecules. This large number of molecules makes it likely that a photon travelling through the outer segment will undergo absorption and trigger the visual process. At the opposite end of the photoreceptor is the synaptic ending, which informs bipolar and horizontal cells about the light being absorbed in the outer segment. News of the light is carried from the outer segment to the synaptic ending by an electrical signal, a change in the voltage across the surface membrane. How is this signal generated, and what does it say?

The pigments that initiate vision are members of a broad class of receptor molecules that sense signals originating outside cells. All such molecules share a structural motif in which a single chain of amino acids is embedded in a lipid membrane (Figure 3). The amino acid chain is folded into a roughly cylindrical structure and traverses the membrane seven times. Light is absorbed by an accessory group buried in the centre of the pigment molecule. This group, the chromophore, consists of 11-*cis* retinal, an isomer of vitamin A aldehyde in which the side chain is bent at the 11–12 carbon bond in the side chain. At its aldehyde end the chromophore is covalently linked to the seventh transmembrane segment in the amino acid chain. Absorption of a photon by retinal allows a rotation to occur at the 11–12 carbon bond in the chromophore, causing the side chain to straighten. This event, *cis–trans* isomerisation, is the only direct effect that light has in vision.

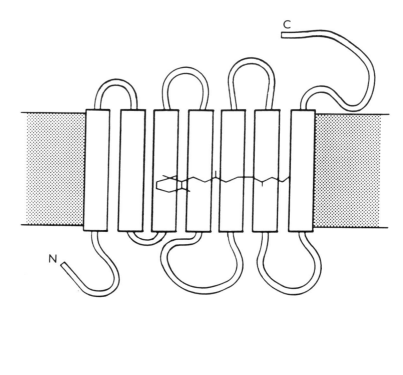

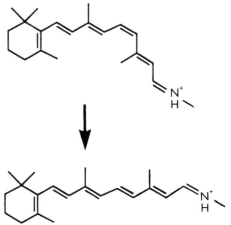

Figure 3 The pigment rhodopsin and its chromophore. *Top:* The protein, consisting of a single chain of amino acids, with transmembrane segments shown as open rectangles. The protein is embedded in a lipid membrane (shown dotted) within the outer segment. The chromophore, 11-*cis* retinal, lies between the membrane-spanning segments. *Below:* How light absorption (arrow) changes the shape of the chromophore's carbon skeleton.

When the side chain of retinal straightens, the chromophore pushes on the protein and changes its three-dimensional shape. The new protein structure has catalytic activity. In the new state, the pigment molecule activates another protein called transducin. This process is catalytic: active pigment is not consumed, but instead provides a template on which a large number of transducins are serially activated. Hundreds of active transducin molecules are formed within a fraction of a second by a single active pigment molecule, so that large amplification is achieved immediately. Activated transducin turns on a third protein, a phosphodiesterase. This enzyme breaks down a diffusible messenger substance, cyclic GMP (cGMP), inside the outer segment (Figure 4). The messenger is present at relatively high concentrations in darkness and binds to specialised sites in the surface membrane. When one of these sites binds cGMP, a tiny hole opens through the membrane. Positively charged ions, mainly sodium, flow into the cell through the open holes, propelled by diffusion and the voltage difference across the membrane. The entry of positive charge prevents the voltage on the inside of the cell becoming as negative as it is in most cells. Light-triggered activity of the phosphodiesterase destroys cGMP, which is no longer available to bind to the pores in the surface membrane. The pores close, and sodium ions can no longer enter. The voltage inside the cell becomes more negative with respect to the outside. In darkness, the interior voltage might typically be -30 mV,

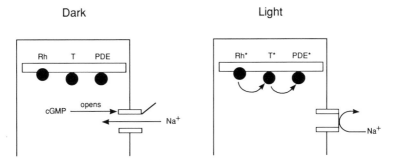

Figure 4 Cyclic GMP (cGMP) cascade of vision. In darkness cGMP opens membrane pores that allow sodium ions (Na$^+$) to enter the outer segment. In the light photoexcited rhodopsin (Rh*) activates transducin (T*), which activates phosphodiesterase (PDE*). The phosphodiesterase breaks cGMP down, and the concentration of cGMP therefore drops in the light. The pores close, and Na$^+$ ions are unable to enter.

and in bright light it approaches −70 mV. Somewhat similar chains of events, triggered by receptor molecules and amplified by analogous enzyme cascades, occur in cells that detect other stimuli such as hormones, odorants and neurotransmitter substances.

The negative swing in the transmembrane voltage spreads over the length of the cell to the synaptic ending. Here, it lowers the rate at which a synaptic transmitter substance, the amino acid glutamate, is secreted onto the bipolar and horizontal cells. These cells sense that light has been absorbed in the rod or cone by detecting a fall in the concentration of glutamate.

Rod vision is colour-blind

A rod or cone informs bipolar and horizontal cells of only one thing: how many photons it has absorbed in the recent past. This information is contained in the size of the voltage swing across its membrane, and the size of the consequent reduction in glutamate release. A larger signal means more absorbed photons because photons contribute to the signal in an additive fashion. The number of photons that the cell absorbs is determined by two factors: the wavelength and the intensity of the incident light. The wavelength determines the likelihood that an incident photon will be absorbed. Once absorbed, however, a photon of any wavelength causes the same *cis–trans* isomerisation of the retinal chromophore and the same electrical response. At Stanford University in 1979, King-Wai Yau, Trevor Lamb and I verified that the single photon response was independent of wavelength. We drew a single outer segment of a rod into a close-fitting pipette (Figure 5) connected to a current-measuring amplifier, and we stimulated it with flashes so dim that on average only one photon was absorbed per flash. Single photons of long, middle and short wavelength elicited identical responses. While the wavelength fixes the probability that a photon will be absorbed, the intensity fixes the number of photons that have a chance to be absorbed. The number absorbed depends upon both variables.

This state of affairs makes it impossible for one cell, rod or cone, to signal separately wavelength and intensity. Consider a rod upon which fall 100 photons at 500 nm wavelength. These photons will be absorbed

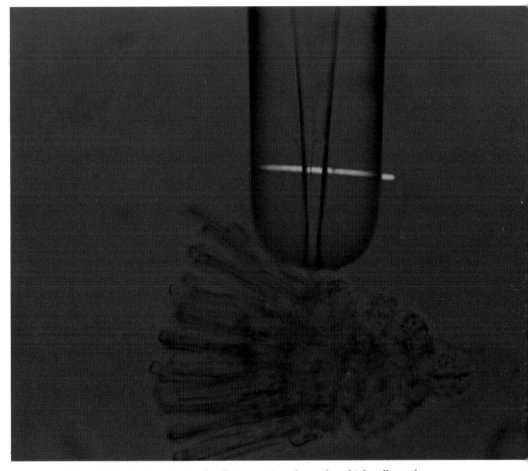

Figure 5 A single retinal rod cell in a suction electrode, which collects the cell's membrane current. A slit of light about 2 μm wide is imaged on the outer segment. This method allowed the single-photon response to be recorded and studied at different wavelengths.

with a probability of say 10%, so that a total of ten absorptions will occur. Ten absorptions would also occur if 1000 photons were incident at 600 nm, a wavelength where the probability of absorption is only 1%. Since the cell reports only the number of photons absorbed, the signals generated by the two lights are identical, even though their wavelengths are very different – no colour (wavelength) information is available. This explains why in starlight, when only the rods contribute to vision, we have no colour sensation.

How three types of cone cell encode wavelength

Strong indirect evidence for Young's suggestion of three resonators in the eye came from work in the last century by Maxwell and Helmholtz, who analysed colour matching experiments. It was found that a suitable mixture of three monochromatic lights in the blue, green and red part of the spectrum could duplicate the colour sensation produced by light of any arbitrary spectral composition. Today colour television sets put this result to practical use. The appropriate mixture of light from the red, green and blue guns produces the desired colour appearance at each point in an image. Maxwell and Helmholtz suggested that the minimum of three primary lights needed to generate all colour appearances is the number of Young resonators in the eye.

Recast in modern terms, their notion was that wavelength is specified by the relative sizes of the responses of three types of cone, each containing a pigment (Young resonator) that preferentially absorbs light in one region of the spectrum – short, middle, or long wavelength. For simplicity, I call these three types of cone blue, green and red, respectively. A monochromatic light at a given wavelength would be absorbed to different extents by the three types of cell, each of which would generate a response proportional to the number of photons it absorbed. The relative sizes of the three responses would specify wavelength, and the brain could infer this wavelength by measuring the relative sizes of the cone responses. A monochromatic test light and a matching mixture of three monochromatic primary lights would appear identical if the test light and mixture elicited identical cone responses.

Elegant as these ideas were, they left the resonators undefined. In the mid 1960s a more direct approach was taken by P. K. Brown and G. Wald at Harvard University and W. B. Marks, W. H. Dobelle and E. F. MacNichol at Johns Hopkins University. They shone a small beam of light across a single cone on the stage of a microscope and measured its ability to absorb light as a function of the wavelength. They found three types of cones, which absorbed in the blue, green or red part of the spectrum.

More recently, measurements of this type have been beautifully refined by J. K. Bowmaker, J. A. Dartnall and J. D. Mollon, at Sussex University, Queen Mary College, London, and the University of Cambridge.

In 1987, the late Brian Nunn, Julie Schnapf and I made measurements

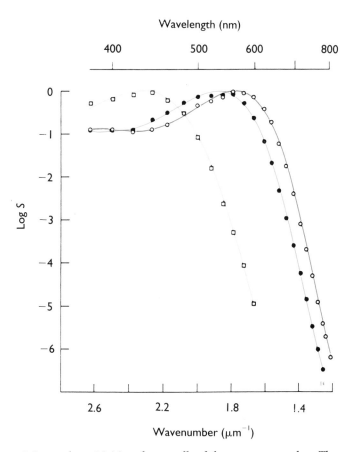

Figure 6 Spectral sensitivities of cone cells of the macaque monkey. The *y*-axis shows the relative sensitivity to a quantum of light on a logarithmic scale. The *x*-axis is wavelength (upper scale) and its reciprocal, wavenumber (lower scale).

over a wider range by using the method illustrated in Figure 5. We recorded a single cone's electrical responses and used these to determine the cell's sensitivity to monochromatic lights throughout the visible spectrum. We term this property the *spectral sensitivity* of the cell. The cones were from the eyes of macaque monkeys that were the donor animals in heart–lung transplant experiments. The points in Figure 6 summarise the spectral sensitivities of forty-one cones. Five cells were blue, with peak sensitivity near 450 nm. Twenty cells were green, with peak sensitivity near 530 nm, and sixteen cells were red, with peak sensitivity near 560 nm. We found that the spectral sensitivities of red and green cones from a human eye that had to be removed because of a tumour were virtually identical with those of the respective monkey cones. Since the ordinate scale in Figure 6 is logarithmic, the vertical displacements between the curves at any wavelength give the factors by which the excitations of the three cones differ when they are stimulated by monochromatic light at that wavelength.

Does one type of cone cell contain a mixture of pigments with one predominating, or is only one pigment present? The measured spectral sensitivities indicate that there is only one. In Figure 6 the blue cone curve at 600 nm lies five log units below the green and red curves. This indicates that, in a blue cone, less than one pigment molecule in 100 000 is of the red or green type. It will be fascinating to learn how the expression of the genes for the pigments is regulated stringently enough to produce this great purity, which is desirable for optimally registering wavelength. Suppose instead, that all cones contained equal amounts of all three pigments. Even though the three pigments would be excited to different extents by lights of different wavelength, all cells would respond equally and no wavelength information whatsoever would be available in their signals. In the less extreme case where each type of cone contained predominately one pigment, molecules of the 'wrong' type would blur the differential cone excitations and impair the ability to sense small differences in wavelength.

The spectral sensitivities of the three types of cone provide a physiological basis for several kinds of psychophysical observations on human colour vision. One example is the colour matching experiment performed by W. S. Stiles and J. M. Burch in the 1950s. A subject viewed a screen on

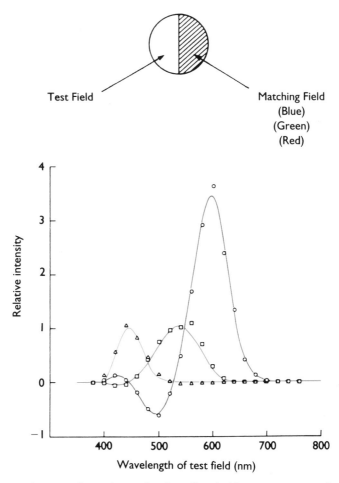

Figure 7 Colour-matching, observed and predicted. *Above*: arrangement for experiments. A monochromatic test light of selectable wavelength was projected in the left field. On the right were three monochromatic matching lights of fixed wavelength. *Below*: intensity of the matching lights as a function of the wavelength of the test light. Continuous curves: Stiles and Burch's results. Points: results predicted from monkey cone spectral sensitivities given in Figure 6. A dip below zero indicates that the matching light had to be applied to the left rather than the right half of the field.

which was projected a divided field of light (see top of Figure 7). In the left half of the field there was a monochromatic light of standard intensity at a test wavelength selected by the experimenter. In the right half of the field there was a mixture of three monochromatic primary lights at fixed wavelengths in the red, green and blue regions of the spectrum. After the experimenter selected a test wavelength, the subject adjusted the intensities of the three matching lights until both halves of the field appeared identical. When the match was made, each cone system was identically stimulated in both halves of the field. Stiles determined the rules that govern how the intensities of the matching lights depend on the wavelength of the test light. He could not infer from these rules the spectral sensitivities of the three types of cone. Knowing the cone sensitivities from the experiment illustrated in Figure 6, however, we can work forward and attempt to predict Stiles' rules. The lower panel in Figure 7 illustrates that the prediction works. This correspondence and others indicate that we have a satisfactory quantitative description of how colour information is represented at the level of the cones.

Molecular structure and genetics of cone pigments

In 1986 Jeremy Nathans, a medical student at Stanford, isolated and identified the genes for the three cone pigments of humans. He isolated the genes by assuming that they would be structurally similar to the rhodopsin gene, which he had isolated earlier. The similarity caused the rhodopsin gene to adhere to the cone pigment genes, allowing them to be 'fished out' of the genome. The DNA base codes of the three genes were translated to sequences of amino acids in the proteins for which they coded, revealing that the three cone pigments are strikingly similar to rhodopsin. The structures are shown schematically in Figure 8. Like rhodopsin, each cone pigment consists of a single long chain of about 350 amino acid residues. A sizeable fraction, roughly 40%, of the amino acids in rhodopsin and the cone pigments are identical, as indicated by the coloured circles in the upper two panels of Figure 8. Analysis indicated that the cone pigments, like rhodopsin, contain seven membrane-spanning regions. The red and green pigments have amino acid sequences that are 96% identical (lower right panel). Apparently they diverged relatively recently in evolution.

How do we know which gene codes for which pigment? Nathans found that one gene was located on chromosome seven, and the other two were located very near one another on the X (sex) chromosome. Since it was

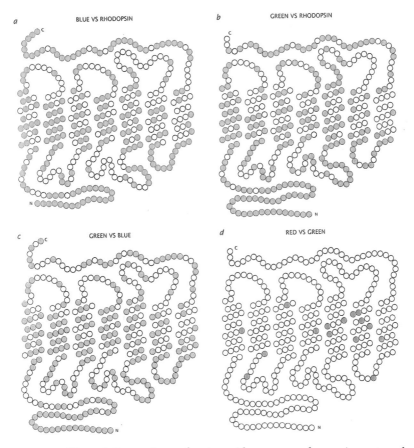

Figure 8 Comparisons of amino acid sequences of cone pigments and rhodopsin. Each circle denotes one amino acid residue. Filled circles indicate differences in amino acid residues. The polypeptide chains are shown oriented as in Figure 3.

well known that defects in red–green color vision are usually sex-linked, the gene on chromosome seven presumably coded for the blue pigment. The genes on the sex chromosome were identified by examining genes of males on whom colour testing had revealed that the red or green pigment was lacking. More recently, Merbs and Nathans have clinched the identification by remarkable experiments in which genetic messages for each pigment were inserted into kidney cells kept in tissue culture. A given message caused the cells to synthesise a cone pigment whose absorption at different wavelengths could be measured in a test tube and thus characterised directly as red, green or blue.

Nathans' analysis of the red and green gene locus on the X-chromosome gave a plausible mechanism for several inherited defects of colour vision, including the red–green 'colour blindness' first described by the chemist John Dalton in the late eighteenth century. The two genes are located in a head-to-tail arrangement on the X-chromosome and are almost identical. This makes them susceptible to erroneous duplication when the genetic material is prepared for packaging into egg and sperm cells during meiosis. In this process the paired strands of DNA in a chromosome are cut by an enzyme and the cut ends of the different strands are cross-connected. This shuffles the genetic cards that are loaded into the germ cells. The strong similarity of the red and green genes' coding sequences increases the chance that a cut will be made at the wrong place. Erroneous cuts generate the flawed arrangements of the red and green genes illustrated in Figure 9. Since females have two X-chromo-

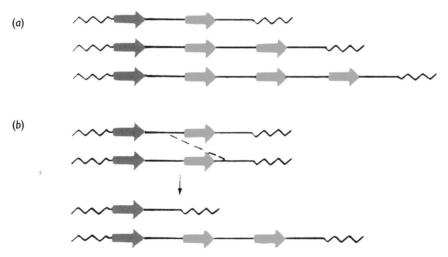

Figure 9 (*a*) Three normal variants of the arrangement of red and green genes on the X-chromosome (zigzag line). (*b*) The dashed line indicates an erroneous recombination during meiosis. This causes the green gene to be deleted from one chromosome and an extra copy of the green gene to be put on the other chromosome. Some errors cause the formation of hybrid genes (not shown).

somes, only one is likely to be flawed and the correct one will work normally.

Males, who have only a single X-chromosome, are hit harder by incorrect shuffling. When the green gene is deleted, the green cone pigment is absent, and the individual has little information about colour in the long wavelength end of the spectrum. Sometimes what should be the green gene is a hybrid red/green gene that codes for a pigment that behaves as red; again the green pigment is lacking. Lack of the red pigment occurs if a hybrid red/green gene is made and codes for a pigment that behaves as a green pigment. Individuals lacking the red pigment also have impaired colour discrimination at long wavelength. For someone lacking either the red or the green pigment, only one pigment operates at long wavelength, and essentially no colour information is present in this region of the spectrum. Monochromatic yellow light at 580 nm, for instance, could be matched by green light at 540 nm or red light at 620 nm. Some hybrid genes direct synthesis of pigments that absorb at wavelengths intermediate between those of the normal red and green pigments. Individuals with these genes may have three pigments, but will make colour matches different from those made by the majority of trichromats (see Mollon, Chapter 5).

Why do the cone pigments absorb selectively in the blue, green or red part of the spectrum? In all three pigments (as well as rhodopsin) the absorber of light is the same: 11-*cis* retinal. It is the proteins that confer individuality by modifying the particular wavelengths that retinal absorbs. They do so by influencing the distribution of electrons in retinal. It is these electrons that interact with the incident light wave and, when absorption occurs, capture its energy. If the protein causes the electrons in retinal to become more delocalised (able to move freely over the chain of double and single bonds between carbon atoms), light at longer wavelengths will be preferentially absorbed. The protein in the red cone pigment exerts the strongest delocalising effect, the protein in the blue cone pigment the least.

Two hypotheses were advanced to explain how the proteins might cause different amounts of electron delocalization. Barry Honig and his colleagues suggested that the protein contains negative charges that are located near the electrons that form chemical bonds between the carbon

atoms in the side chain of the retinal molecule. By promoting delocalization these charges would cause light of longer wavelength to be absorbed.
A second notion, proposed by Richard Mathies, was that the protein
might twist the retinal molecule in regions where the bonding electrons
are distributed. Clint Makino, Tim Kraft and I were able to reject the
possibility that twist between the ring and side chain of retinal explained
the different behaviour of the red and green cone pigments. We replaced
the chromophore with a modified form of retinal that could not be
twisted at this bond. Had red and green cone pigments behaved differently because the two proteins twisted the chromophore differently, the
pigments should have absorbed identically when they contained the modified retinal. This did not occur – the difference in the absorptions of the
two pigments persisted. It now appears that differences in the number
and positions of negative charges are mainly responsible for producing
the different spectral absorptions of the visual pigments. The different
behaviour of the red and green cone pigments is explained by the presence in the red pigment of three amino acids that contain hydroxyl
groups whose oxygen atoms behave as weak negative charges.

What the eye tells the brain

As Figure 2 shows, signals generated by the cones are relayed to bipolar
cells and thence to retinal ganglion cells. The ganglion cells transmit
visual information over the optic nerve to the brain. They do so with
nerve impulses – large, brief signals that relay information over long
nerve fibres throughout the body. What exactly do a given cell's impulses
mean? In 1953 Steven Kuffler and Horace Barlow approached this question. They used a microelectrode to record impulses from a single ganglion cell while stimulating the retinal surface with light. The cell gave
impulses even in darkness, but application of light to the retinal surface
changed the number of impulses generated per second. Visual signals
thus consisted of increases or decreases in the frequency of impulse generation. A given cell responded only to illumination of a localised region
of the retinal surface, corresponding to a localised region of the eye's field
of view. Typically this region of retinal surface, the receptive field, consisted of two concentric regions that had opposite effects on the rate of

impulse generation. The two regions are shown schematically in the upper left-hand diagram in Figure 10, in which the plane of the page lies in the plane of the retinal surface.

For this example, light applied in the circular central region causes impulses at higher rate (denoted by the plus (+) sign), while light applied in an annular surrounding region suppresses impulses (denoted by minus (–) sign). This cell responds most strongly when the central region is illuminated and the surrounding region is in darkness. Apparently impulses in this cell indicate the presence of a small spot of light on a dark background. Another type of cell (not shown) responds best to a small dark spot on a bright background. Both types of cell signal local contrast in the retinal illumination. Local contrast is important because it delineates the borders of objects in the visual world.

Information about colour is transmitted by a set of ganglion cells

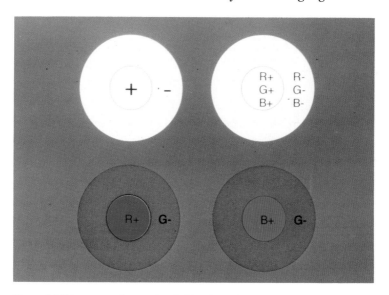

Figure 10 Diagrams of receptive fields of ganglion cells, with the surface of the retina in the plane of the paper. Plus signs (+) indicate regions in which light causes responses; minus signs (–) indicate regions where light suppresses responses. Upper left panel: one type of field described by Kuffler and Barlow. Upper right panel: a black-white contrast field described by Wiesel and Hubel. Stimulation of all three types of cone in the central area causes responses, while stimulation in the surround suppresses responses. Lower panels: two colour-selective receptive fields.

different from those that encode black/white contrast. This was demonstrated nearly thirty years ago by T. N. Wiesel and D. H. Hubel, who examined the receptive fields of single neurons in the lateral geniculate nucleus of the monkey brain, extending the analysis to include the effects of wavelength. Cells in this nucleus have receptive fields that are virtually identical with those of the retinal ganglion cells that drive them, so that the recordings could equally well have been made from ganglion cells.

The upper right panel of Figure 10 shows a typical receptive field of a cell that signals black/white contrast. It consists of concentric antagonistic regions of the sort described by Barlow and Kuffler. Within each region, signals from all three types of cone are added together. Stimulation of the central region with white light, which stimulates all three types of cone, evokes impulses, but so also does monochromatic light that selectively stimulates only one type of cone. Similarly, red, green, blue or white light on the surrounding region suppresses impulses (minus signs).

Another class of cell transmits wavelength information. The lower left panel in Figure 10 illustrates one common variety of receptive field. Again there are concentric regions with antagonistic effects on the rate of firing impulses. The central area, which evokes impulses, however, behaves as if it receives only from red cones; although blue and green cones are present, only signals from the red cones reach the ganglion cell. For stimuli confined to the central region, red light, which selectively stimulates red cones, is more effective than green or blue light. The annular surround suppresses impulses when it is illuminated, and receives only from green cones. Monochromatic light in the green part of the spectrum is therefore most effective in suppressing impulses. If the receptive field is illuminated with diffuse (uniform) monochromatic light, firing is accelerated by light of long wavelength, which preferentially stimulates the red cones in the central region, and firing is inhibited by light of shorter wavelength, which preferentially stimulates green cones in the surround. In diffuse light, then, this cell compares the activity of the red and green cones, a first step in extracting wavelength information. A somewhat similar type of cell compares the activity of blue and green cones (lower right panel in Figure 10).

The cellular interactions that give rise to the colour-coded receptive fields of retinal ganglion cells are still not completely understood. It is clear, however, that they require highly selective synaptic connections between retinal neurons. The centre of the receptive field probably depends upon colour-specific bipolar cells that synapse upon the ganglion cell. Such bipolar cells collect signals from only one of the three types of cone. The receptive field surround may depend upon specific connections between horizontal cells, cones and bipolar cells.

The notion that different kinds of information, say colour and spatial contrast, are processed by different groups of cells has been supported by anatomical findings as well as physiological experiments. For example, within the lateral geniculate nucleus of the brain, the ganglion cells that encode black/white contrast terminate in regions that are spatially separated from the regions in which the wavelength-selective cells terminate. Clinical observations also indicate that colour is processed in regions of the cerebral cortex different from those that handle other types of visual information. Some patients with localised cortical lesions have severely impaired colour vision, even though the retinal cones are normal and the ability to perceive form and motion is normal. Recently, Semir Zeki and his colleagues have used positron emission tomography (PET) scans, which detect regional changes in blood flow resulting from neural activity, to confirm the existence of a colour-processing centre on the undersurface of the back of the human brain.

Perhaps surprisingly, most wavelength-selective ganglion cells do not transmit unambiguous information about wavelength. Instead, the majority carry mixed messages about the wavelength and the spatial distribution of the stimulus. The ambiguity is somewhat reminiscent of that between wavelength and intensity in a single cone cell. Consider for example the receptive field shown at the lower left of Figure 10. For spatially uniform light the cell compares the outputs of red and green cones: monochromatic light at the red end of the spectrum, which preferentially excites red cones, causes impulses, and light in the green part of the spectrum, which preferentially excites green cones, suppresses impulses. For white light, however, the cell registers spatial contrast. Thus, if white light is applied only to the centre of the receptive field it will excite the red cones and thus trigger impulses in the ganglion cell. White light applied only to the annular surround will stimulate the green cones in the

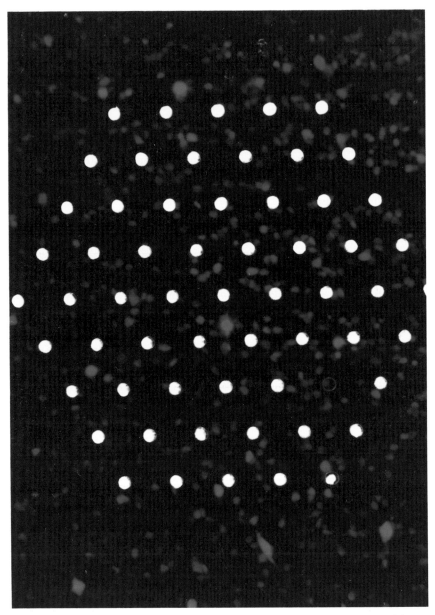

Figure 11 A toad retina and the multielectrode array used to record from retinal ganglion cells. The platinum discs that pick up impulse activity are shown white. Each is 15 μm in diameter. The cell bodies of retinal ganglion cells appear orange; they were stained by a fluorescent label.

surround and thus suppress impulses. For white light, therefore, the best stimulus is a small spot on a dark background. An impulse in the cell thus means, roughly, that there is either a small white spot on a dark background or a spatially uniform light that contains predominately long wavelengths. Spatial and chromatic information are ambiguous. Furthermore, one cell does not compare the activity of all three types of cone, but only two of them.

These considerations indicate that, at the level of the retinal ganglion cells, just as at the level of the cones, wavelength information is not available in the activity of any one cell. Instead groups of ganglion cells carry colour signals from any point on the retinal surface. In order to extract wavelength information, the outputs of different types of ganglion cell need to be pooled and compared.

How should the pooling and comparison be done? What does a multi-neuronal colour signal in the optic nerve look like? In order to approach these and related questions, Markus Meister (now at Harvard) and I have begun to examine the activity of groups of retinal ganglion cells. A piece of isolated retina is placed on a planar multielectrode array consisting of 61 tiny platinum discs spaced evenly over a region about half a millimetre in diameter (Figure 11). Typically each disc picks up impulses from a few ganglion cells, so that recordings can be made from a population of up to 100 cells. Light stimuli are applied by imaging, onto the retina, the screen of a computer monitor that displays any desired spatial, chromatic or temporal pattern. This approach promises to give interesting new information about multineuronal colour signals.

Conclusion

Great progress has been made in unravelling the eye's mechanisms for colour reception, yet fascinating questions remain. How, for example, does a single cone cell decide which pigment to express? How can the expression be regulated so very stringently while preserving the flexibility to avoid 'holes' in the cone mosaic when the gene for a particular pigment is missing? What signals direct the formation of colour-specific bipolar and ganglion cell channels? How do multineuronal signals in the optic nerve encode visual stimuli, and how does the brain decode them?

Surely Thomas Young would have been fascinated at our deepening insights into his three resonators and the mechanisms that the nervous system uses to measure their activity.

Acknowledgements

I thank Drs Leon Lagnado and Clint Makino for reading the manuscript. The research described here was supported by grants from the National Eye Institute, Retina Research Foundation and the Alcon Research Institute.

Further reading

Boynton, R. M., *Human Color Vision*, New York: Holt, Rinehart and Winston, 1979.

Hubel, D. H., *Eye, Brain and Vision*, Scientific American Library series, no. 22, New York: Scientific American, Inc., distributed by W. H. Freeman, 1988.

MacAdam, D. L. (ed.), *Sources of Colour Science*, Cambridge, MA: MIT Press, 1970.

Nathans, J., 'The genes for colour vision', *Scientific American*, **260** (1989), 42–9.

Nicholls, J. G., Martin, A. R., and Wallace, B. G., *From Neuron to Brain*, Sunderland, MA: Sinauer Associates, Inc., 1992.

Schnapf, J. L., and Baylor, D. A., 'How photoreceptor cells respond to light', *Scientific American*, **256** (1987), 40–7.

Stryer, L., 'The molecules of visual excitation', *Scientific American*, **255** (1987), 42–70.

5   Seeing Colour

John Mollon

Figure 1 Christine Ladd-Franklin (1847–1930), pioneer American feminist,
who put forward an evolutionary account of human colour vision.
Ferdinand Hamburger, Jr, Archives of the Johns Hopkins University.

Even the Preface to her collected papers describes her as belligerent. She
graduated from Vassar while Darwin was still in his prime, and to the

time of her death in 1930 she robustly defended the evolutionary theory of colour perception that she first put forward in 1892. Without the advantage of molecular biology, Mrs Christine Ladd-Franklin (Figure 1) surmised that the light-absorbing molecule of the retina had undergone successive differentiations, to give first achromatic, then dichromatic, and finally trichromatic vision. I believe, with Mrs Ladd-Franklin, that to understand our own colour vision we must understand how it came to be the way it is. I shall argue that it depends on two distinct subsystems, a relatively recent one overlaid on a phylogenetically ancient one. Denis Baylor (Chapter 4) has already introduced the idea that our rich experience of hue depends on the proportions of light absorbed in just three classes of cone cell in our retinae (see Figure 2(c)). What I should like to do here is to emphasise that the three types of cone are not equal members of a trichromatic scheme, but rather that they evolved at different times for different purposes.

The ancient subsystem of colour vision

For most diurnal mammals the main business of vision (the discrimination of position, form and movement) depends on a single class of cone with its peak sensitivity in the middle of the spectrum. Recall, however, from Chapter 4 that any individual class of cone is colour blind and so we need to be able to compare the photon absorptions in at least two classes of cone before we can discriminate colour. A long time ago – probably before the mammals evolved – a second population of cones, sensitive to short-wave (violet) light, were added to the array of light-sensitive cells: a comparison of their signals with those of the predominant class of cones provided the ancient system of colour vision (Figure 2(a)). In the typical mammalian retina, these violet-sensitive cones are sparsely scattered within the light-sensitive array, and in primates and human beings they represent only a tiny minority of all cones. Using the technique of microspectrophotometry, in which a 2 μm wide beam of light is passed through individual cone cells, James Bowmaker and I have recently been able to measure the absorption spectrum of each cone in small patches of intact retina. In the central region of the retina, the fovea, we find that only about 3% of all cones are of the short-wave kind,

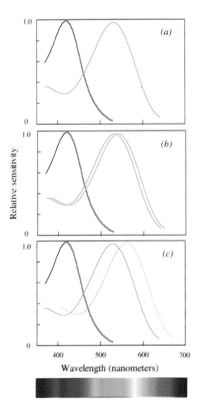

Figure 2 The light absorption curves of pigments in the cone cells of the eye. Each panel shows the relative proportion of light absorbed by a given photopigment at different wavelengths.

(a) Dichromacy. This form of colour vision is found in many non-primate mammals, in male New World monkeys, and in about 2% of human males. It depends on comparing the relative light absorption in violet-sensitive cones with the absorption in more numerous cones that have peak sensitivity in the green or green–yellow part of the spectrum.

(b) Anomalous trichromacy. This form of colour vision is found in about 6% of human males: a limited discrimination is achieved in the red–green range by comparing the absorptions in two pigments that have very similar positions in the spectrum. Some female New World monkeys exhibit similar sets of photopigments.

(c) Normal trichromacy. This panel shows the arrangement of three photopigments found in Old World monkeys, in apes, and in those human subjects with normal vision. The exact spectral positions of the pigments vary slightly between species and between individuals.

and in a small area at the very centre they may be completely absent.

So the ancient mammalian system of colour discrimination depends on a comparison of the rates at which photons are absorbed in the sparse short-wave cones and in the more numerous type of cone that has its peak sensitivity in the middle of the spectrum. The comparison is carried out by nerve cells within the retina that draw inputs of opposite sign – excitatory or inhibitory – from the two classes of cone cell (Figure 3(a)). It is this subsystem of colour vision that is commonly preserved in those of our own species whom we label 'colour-blind'. Phenomenologically, in the case of both the colour-blind and the colour normal, the ancient subsystem divides colours into 'warm' and 'cold': stimuli appear warm if long wavelengths predominate and cold if short wavelengths predominate. A mother will find that she can communicate unambiguously with her colour-blind son if she speaks only in terms of this warm–cold dimension.

Why are the short-wave cones so sparse? The answer is probably chromatic aberration, one of the several optical imperfections of the eye. Von Helmholtz once said that, if an instrument maker sent him the human eye, he would send it back. (There are actually rather few customer complaints, and we may guess that this is because the retinal image is the one optical image never intended to be looked at; for the brain's purpose is to reconstruct the *external* world, not to show us the distorted and degraded image on the retina.) Chromatic aberration arises from the fact that blue rays are more refrangible than red. The two ends of the spectrum cannot be concurrently focused by the human eye. The eye normally focuses for the yellow light that is so abundant in our world and so the short-wave component of the retinal image is permanently out of focus. The visual system would gain little by sampling the blue and violet component of the image at higher density. The Great Instrument Maker provides us with short-wave cones primarily for colour vision. We make little use of them for many of the functions of the eye, such as the discrimination of fine detail or the detection of movement and flicker.

It is easy to demonstrate the poor spatial resolution of the ancient subsystem of colour vision. Take a blank transparency for an overhead projector and, with a suitable felt-tipped pen, tint one half of it a pale lime yellow (Figure 4). The boundary of the yellow-coloured area should be sharp and straight. When the transparency is projected with an overhead

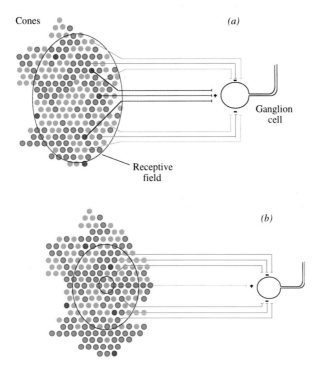

Figure 3 Ancient and modern subsystems of colour vision. (*a*) The phylogenetically older subsystem. A retinal ganglion cell draws excitatory (+) inputs from the short-wave cones and inhibitory (–) inputs from the long- and middle-wave cones in a small local region of the photoreceptor array. When this small region (the receptive field of the ganglion cell) is illuminated with homogeneous white light, the ganglion cell will not respond, since its excitatory and inhibitory inputs are in balance; but it will respond vigorously when blue or violet light falls on the receptive field. Short-, middle- and long-wave cones are represented by violet, green and yellow discs, respectively, and the spatial distribution of the different types follows that of Mollon and Bowmaker.

(*b*) The phylogenetically recent subsystem. A ganglion cell draws opposed inputs from the long- and middle-wave cones. The excitatory and inhibitory areas of the receptive field are spatially concentric, and in the foveal region of the retina the centre input to the ganglion cell may be drawn from only a single cone.

The reader should envisage that the photoreceptor array is analysed in parallel by many ganglion cells of the two types, each cell having its local receptive field. References to 'subsystem' in the main text refer to all the cells of a given type.

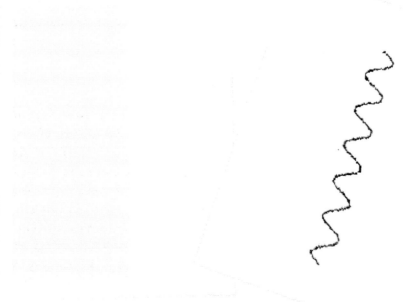

Figure 4 Two transparencies for demonstrating the poor spatial resolution of the older subsystem of colour vision. When the two are superposed using an overhead projector, with the sinusoidal line coinciding with the central edge of the yellow–green region, the observer sees the colour follow the concavities and convexities of the superposed black line. The yellow–green should be as light as possible so as to minimise the brightness difference at its edge. This demonstration was designed by R. M. Boynton.

projector, the edge between the two areas will be visible mainly to the short-wave, violet-sensitive cones: these cones respond more strongly to the white region than to the yellow, whereas the other cones signal little difference between the two regions. Now prepare a second transparency with a sinusoidal black line running down the middle. When this wavy line is superposed on the yellow-white edge, the colours will appear to follow the black line, the yellow spreading into the positive-going excursions to the one side and the white passing into the positive-going excursions to the other. The colour signals from the short-wave cones give only inexact information about spatial position, and it seems that signals from the more abundant longer-wavelength cones are used to decide the exact position of an edge.

The modern subsystem of colour vision

About thirty million years ago, a second subsystem of colour vision evolved in our primate ancestors. This development, like so much of evolution, arose through the duplication and modification of an existing gene. The ancestral gene was located on the X-chromosome and it coded for the light-sensitive pigment found in the main class of mammalian cone, having peak sensitivity in the middle of the spectrum. It now gave rise to two genes, with slightly different DNA sequences, which coded for two photopigments with peak sensitivities in the green and yellow-green regions of the spectrum (Figure 2(c)). The pigment with peak sensitivity in the green is conventionally known as the 'middle-wave' pigment; that with a peak in the yellow-green is known as the 'long-wave' pigment. The two pigments are segregated in different cone cells, and our second subsystem of colour vision depends on a neural comparison of the light absorption in these two kinds of cone. This newer subsystem allows us to make discriminations in the red–yellow–green region of the spectrum, and, because both types of cone are present in high numbers, we enjoy better spatial resolution for this dimension of colour experience. Notice that neither of the pigments has a peak in the red: a red light is simply one that produces a high ratio of long-wave to middle-wave cone signals.

Diurnal birds and many surface-dwelling fish exhibit good colour vision. But amongst the mammals, we share our own full trichromatic vision only with the Old World monkeys and the apes. It increasingly seems that the second subsystem co-evolved with a class of tropical trees characterised by fruits that weigh between 5 and 50 g, are too large to be taken by birds, and are yellow or orange in colour when ripe. The tree offers a signal that is salient only to agents with trichromatic vision. This hypothesis of co-evolution of the tree's signal and the monkey's vision can be traced back to nineteenth-century naturalists, but it takes on new plausibility in the light of recent ecological evidence that there are many species of tropical trees that are dispersed exclusively by monkeys, and these are the trees with yellow or orange fruit (for example, several members of the family Sapotaceae). Conversely, some species of arboreal monkey, such as guenons, rely on fruit for up to 85% of their diet. The tree offers a colour signal that is visible to the monkey against the masking foliage of the forest, and in return the monkey either spits out the undamaged seed at a distance or defecates it together with fertiliser.

In short, monkeys are to coloured fruit what bees are to flowers. With only a little exaggeration, one could say that our trichromatic colour vision – if not the entire primate lineage – is a device invented by certain fruiting trees in order to propagate themselves. Certainly, if we are to understand the regeneration of rain forest, one of the many biological contracts we must understand is that between fruit signals and primate colour vision.

The polymorphic colour vision of New World monkeys

One unsolved puzzle concerns the New World monkeys, which diverged from our own ancestors some 30 million years ago and which have found their own, very remarkable, route to trichromacy. Within a single species of New World monkey, such as the squirrel monkey or the marmoset, colour vision is polymorphic, that is to say there exist in the population several genetically different forms, apparently in stable equilibrium. There may be three types of male and six types of female within a species. The males are always dichromats, resembling colour-blind men in having only two classes of cone (Figure 2(a)) and differing only in the exact spectral position of the pigment with the longer wavelength, whereas a majority of females are trichromats, able to discriminate well in the red–green region of the spectrum. The genetic basis of this variation is instructive and may have analogues in other physiological systems. Like most mammals, the New World monkeys have only a single X-chromosome locus for a pigment in the midde-/long-wave region, but three or more versions of the gene are present in equilibrium in the population, the different versions differing slightly in their DNA sequence and coding for pigments that peak at different positions in the spectrum. A female monkey, having two X-chromosomes, has a good chance of being heterozygous, that is inheriting different versions of the gene from her two parents; since only one X-chromosome is expressed in any given cell of a female mammal's body, she will thus come to have three types of cone in her retina – the two types of cone with peak sensitivity in the green–yellow range plus the violet-sensitive cones that she shares with her male conspecifics. Apparently the visual system of such heterozygous females is plastic enough to take advantage of the signals from the extra cones so as to gain behavioural trichromacy.

This is a particularly pure case of what population geneticists call 'heterozygous advantage': it matters only that the female inherits different versions of the gene from her parents, not which particular ones she inherits. (A recent report suggests that heterozygosity for a related molecule – one of the receptors for dopamine – is associated with reduced susceptibility to schizophrenia.) And there is a moral for those who study the inheritance of traits such as intelligence: here we have a trait, the ability to discriminate red from green, where the variance between individuals is almost all of genetic origin and yet we should expect no correlation between the abilities of parents and offspring. Needless to say, such a remarkable paradox does need some technical qualification. For the statement to be true, it is necessary first that the different versions of the gene are equally frequent in the population; and, secondly, that mating must be random with respect to this gene (for example, females must not be able to distinguish, and favour, males carrying a different version of the gene.)

Traditional field studies of New World monkeys had not given any hint of their polymorphism of colour vision, which emerged from laboratory studies in the 1980s. But the laboratory studies must now feed back to research in the rainforest, for only ecologists in the field are likely to discover why the striking variations between individuals and between sexes are maintained. The biological contract between trees and their primate dispersers is different in the Old World and the New. We do not understand why.

Individual differences in human colour vision

The gene duplication that occurred within the Old World primate lineage remains a source of recombinant mischief for a significant fraction of our own population – those 8% of men who are colour deficient. Almost invariably, it is the newer subsystem that is altered in colour deficiency. The two very similar genes on the X-chromosome have remained juxtaposed, and so are liable to misalign when corresponding chromosomes are brought together at the stage of meiosis during the formation of the ovum. This is the stage when crossing-over occurs; that is, when DNA is exchanged between paired chromosomes. If the chromosomes are locally misaligned and crossing-over occurs in this region, hybrid genes may be formed or a gene may be lost altogether from one of the chromosomes. About 2% of men are dichromats, lacking either the long-wave pigment or the middle-wave pigment. More common, affecting 6% of men, is the milder condition called anomalous trichromacy, in which three pigments are present in the retina, but one of them is displaced in its position in the spectrum (Figure 2(b)).

It is instructive that picking fruit is one of the few tasks in which colour-deficient men are severely handicapped. In one of the earliest reports of colour blindness, in the *Philosophical Transactions of the Royal Society* for 1777, Huddart wrote of the shoemaker Harris, '*Large objects he could see as well as other persons; and even the smaller ones if they were not enveloped in other things, as in the case of cherries among the leaves.*' We especially need colour vision when the background is dappled and variegated; that is, when we cannot use form or lightness to find our target. For urban humans, colour vision is a luxury; but for frugivorous primates, it may be a necessity.

So far, I have referred to normal colour vision as if it were a single condition. But of especial interest is the recent realisation that the modern subsystem varies amongst those of us whose colour vision is nominally normal. You and I may live out our lives in slightly different perceptual worlds. Although we may both enjoy colour vision that is officially normal, coloured objects that look alike to you may look distinctly different to me, and those that look different to you may look identical to me. This small but irreducible discrepancy in our sensations can now be traced to the minimal possible genetic difference; that is, to a single nucleotide difference in our X-chromosomes. Let me elaborate.

Denis Baylor (Chapter 4) has described the structure of the protein molecules on which all our vision depends. They are members of the much larger superfamily of heptahelical receptors, which includes the molecules found in the sensory cells of the nose as well as many of the molecules that recognise the chemical signals between nerve cells. Figure 5 represents photopigment molecules embedded in the multiply enfolded membrane of a retinal cone cell. Each molecule consists of seven helices that span the membrane of the cell. By collating the genes and the photo-pigments of individual New World monkeys, it has been possible to

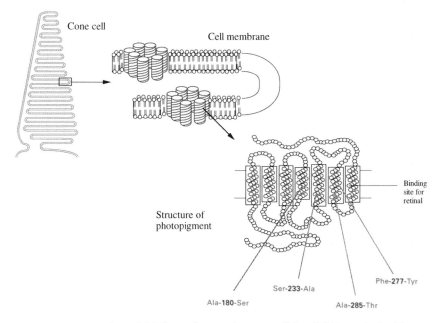

Figure 5 The enfolded membrane of a cone cell (top left) is packed with photopigment molecules. Each of these molecules consists of seven helices, which span the membrane of the cell and are linked by loops outside the membrane. Represented at the bottom right is the sequence of amino acids that makes up this heptahelical molecule. Indicated in the seventh helix is the amino acid (lysine) that binds to the chromophore (retinal) that lends the molecule its spectral absorption. Highlighted in yellow, in the fourth, fifth and sixth helices, are four amino acid residues (numbers 180, 233, 277 and 285 in the sequence) that are known to influence the exact wavelength to which the pigment is maximally sensitive. The codes below represent the alternative amino acid residues at these four positions: in each case the amino acid indicated in green is the alternative that shifts the peak sensitivity of the molecule to shorter wavelengths, that in red the one that shifts sensitivity to longer wavelengths.

identify the surprisingly small number of amino acid residues that make the difference between the long-and middle-wave pigments. Of particular interest is the 180th amino acid residue in the chain, for this is the one that commonly varies in the long-wave pigment in the normal human population. If, on a man's X-chromosome, the long-wave gene codes for the amino acid serine at position 180, then the peak sensitivity of his long-wave pigment will lie at a longer wavelength than if he inherits the code for alanine at this position. Groups of scientists in Seattle and Milwaulkee have shown that this difference is reflected in the settings that different individuals make in a 'Rayleigh match': when men are asked to adjust a mixture of red and green light to match a pure, mono-chromatic, orange light, those with serine at position 180 will tend to choose a mixture with less red in it than will those with alanine. A white male population divides in proportions of about 60:40.

In years to come I think we shall look back on this finding as the first example where a normal variation in our mental worlds is traceable to a polymorphism – a normal variation – in our genes. In this case we know almost all the steps in the chain: we know how the difference of a single nucleotide in the DNA changes the protein that is coded for, we know how that alters the spectral sensitivity of the resulting photopigment, and we know how it will alter neural signals in the visual pathway. Within the next ten years we shall surely know of many cases where differences in our perceptual, cognitive and emotional worlds are traceable to analo-gous genetic variations.

At least with respect to the present issue, women are more compli-cated than men, since they have two X-chromosomes and thus two sets of genes for the long-and middle-wave photopigments.

What happens if – as is statistically very likely – a woman inherits from her parents two different versions of the long-wave gene or two dif-ferent versions of the middle-wave gene? Do the New World monkeys offer us a model? In the case of squirrel monkeys, a basically dichromatic species, the heterozygous females become trichromatic. Is it possible that in our basically *tri*chromatic species, a subset of females become tetra-chromatic, enjoying an extra dimension of colour experience? The best candidates might be women who are carriers of the mild forms of anom-alous trichromacy and whose affected sons must achieve their residual red–green discrimination by means of an abnormal pair of middle-/ long-wave pigments. Hitherto, it has been assumed that such carriers simply share a little in the disability of their sons, exhibiting very mild colour anomalies, rather as the mothers of haemophiliacs have blood that clots more slowly than that of other women. But the mother of an anomalous trichromat must carry at least four genes for cone pigments – the short-wave pigment, and three different pigments in the middle-/ long-wave range. She may for example carry on one of her X-chromosomes the gene for one long-wave pigment and on the other the gene for a different long-wave pigment, one shifted in its position in the spectrum. Owing, then, to X-chromosome inactivation, the retina of such a woman should contain four classes of cone, because the two alternative forms of the middle-wave pigment will be manufactured in different classes of cone. Can she take advantage of them to gain tetrachromatic vision?

Gabriele Jordan and I have been recently made an experimental search for tetrachromatic women. Our subjects included women who were proven carriers (in that they had sons with various forms of colour deficiency and anomaly) as well as women who had normal sons. We used a computer-controlled colour-mixing apparatus to study colour matching in a part of the spectrum where the normal, trichromatic, observer is effectively dichromatic – can make colour matches with only two variables. This is the spectral region from green to red, 550 nm to 690 nm, where the short-wave cones are quite insensitive and where a normal trichromat can match any wavelength with some unique mixture of red and green. What we did was to add an additional variable. We offered our subjects two mixtures. The first mixture (A) was of a red and a yellow and the second (B) was of a green and a deep orange. Each of these mixtures is capable of giving a range of oranges, from a reddish orange to a yellowish orange (Figure 6). The two mixtures were alternated every two seconds and the subject was asked to adjust them until they matched. In successive trials, the computer randomized the starting points for the two mixtures, and the subjects were instructed to find as large a range of matches as they could.

The normal woman, being dichromatic in this spectral region, should be able to make a whole range of matches in the colour space of our

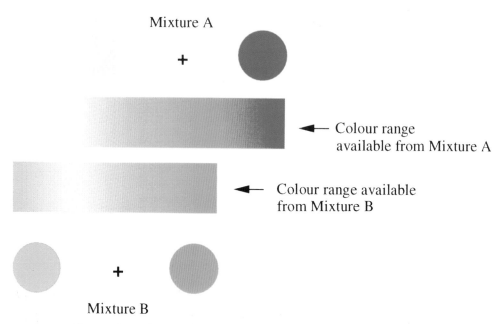

Mixture A

Colour range
available from Mixture A

Colour range available
from Mixture B

Mixture B

Figure 6 A test for tetrachromacy. For the normal observer, Mixture A gives a range of colours from yellow to red, while Mixture B gives a range from green to reddish orange, and so there should be a range of pairs of mixtures that match – mixtures that give colours lying between yellow and deep orange. But for a tetrachromat most of these physically different pairs should look different and only one unique match should be possible.

experiment: there should for her be a range of colours, running from a yellow orange to a reddish orange, that look the same whether they are made by mixing red and yellow or by mixing green and deep orange. But a tetrachromat, having an extra independent signal in this spectral region, would be expected to find only one mixture of the red and yellow that matched a mixture of green and deep orange. In fact, if tetrachromatic women exist, then they are rarer than would be expected from the analogy with New World monkeys. Although our experiments are still in progress and occasional carriers show a very limited range of matches, we are satisfied that the majority of carriers of anomalous trichromacy are not tetrachromats by this test and so the interesting question arises of what it is that allows the female monkey to take advantage of her extra class of cone when most heterozygous women cannot.

Perceptual organisation and the two subsystems

I have argued above that one advantage of colour vision is that it allows us to detect a spectrally distinct target (e.g. a ripe fruit) against a background that varies randomly in lightness and form. A second role for both the ancient and the modern colour systems lies in perceptual organization: colour helps us to identify which elements in a complex visual field belong together, i.e. make up a common object. This capacity of colour to link disparate components of the scene has been well recognised and exploited by painters (see Bomford, Chapter 1, and Riley, Chapter 2), and is today used and misused by those who design screens of information in teletext and other systems.

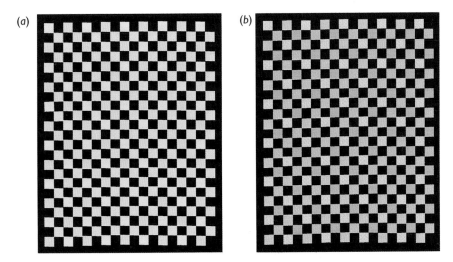

Figure 7 The role of colour in perceptual organisation. In both panels the older colour signal is modulated along the diagonal from lower right to upper left and the newer colour signal is modulated from lower left to upper right. In (*a*), where the modulation of the new signal is greater, the reader with normal colour vision will see the array predominantly running from lower right to upper left, while in (*b*), where the modulation of the older colour signal is predominant, the opposite organisation presents itself. Owing to the limitations of colour reproduction, the chromaticities and luminances shown here do not correspond exactly to those used in the experiments described in the text.

But what is the relative organising power of the two subsystems? Benedict Regan and I have recently been tackling this question experimentally. Our displays look like the arrays in Figure 7. The subject's task is to report whether the predominant organisation of the array runs from lower left to upper right or from lower right to upper left; and we arrange that the old and the new subsystems impose contrary organisations. In Figure 7(a) it is the new subsystem that is dominant, in Figure 7(b) the old. These are abstract arrays, but if we translate the discourse into that of the painter, it will be clear that this is the same organising power of colour as Bridget Riley discusses in Chapter 2.

We vary the signal strength of a given subsystem by varying the colour difference between the alternating lines of the array. In our formal experiments, the computer randomised from presentation to presentation which colour subsystem was associated with which orientation. For a series of colour differences on one of the two chromatic dimensions, the computer adjusted the depth of modulation of the other dimensions to find the point at which on 50% of occasions the subjective organisation was determined by the ancient colour system and on 50% by the new. An important factor that we also varied in the experiment was the spatial separation of the individual patches in the array, from about ten minutes of visual angle to one degree.

Figure 8(a) shows some typical results for a subject with normal colour vision. The vertical axis represents the size of the signal in the ancient colour system, and the horizontal axis, the size of the signal in the modern colour system. The different subsets of data points are for different separations of the individual patches in the array. For any spatial separation, the lines represent the balance points of where the two colour signals have equal organising power; they show how much modulation of the ancient system is equivalent to a given modulation of the modern system. For any one spatial separation, the potency of the two signals grows concomitantly, the data points fall on straight lines. But spatial separation has a dramatic effect: when the patches are close together, a large modulation of the ancient system is required to balance a small modulation of the new system, but when the patches are further apart the ancient colour system, the warm–cold dimension, becomes more potent.

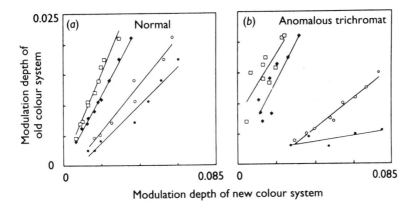

Figure 8 Results for (*a*) a normal observer and (*b*) an anomalous trichromat in a test that measures the relative salience of the signals of the phylogenetically old and new colour systems. Within each panel there are four sets of data points, corresponding to different separations of the elements in the array. Symbols as follows: (□) 0.144° of separation; (◆) 0.181° (○) 0.301°; (●) 0.602°. The ordinates of the graphs are expressed in units of the chromaticity diagram introduced by MacLeod and Boynton (*Journal of the Optical Society of America,* **69** (1979), 1183–6), a diagram whose axes correspond to the two subsystems discussed here. As the separation of elements increases, the older subsystem becomes the more salient, and this effect is exaggerated for the anomalous trichromat.

In other words, a given modulation of the newer system is balanced by a modest modulation of the old. Almost certainly, these results reflect the chromatic aberration of the eye and the sparse distribution of short-wave cones that I described earlier. They have implications for graphic designers and for those who program screens of information for computer displays: if graphic elements or stroke widths are small or are to be discriminated by the public at large viewing distances, then categories of information should not be differentiated by colours that differ mainly in the excitation of the short-wave cones.

Figure 8(*b*) shows analogous results for an anomalous trichromat and give us some insight into his private colour world. When the elements of the array are close together, the relative saliences of the old and new colour signals are similar to normal, but when the elements are well separated the old colour signal is relatively much the more salient and the world of the anomalous trichromat is dominated by the warm–cold dimension.

The central representation of colour

The reader might suppose that the two subsystems I have been describing correspond to the traditional blue–yellow and red–green axes of colour space. There are four hues that to most people appear phenomenally pure or unmixed – red, yellow, green and blue – whereas other hues appear phenomenally to be mixtures: we feel we can see the red and the yellow in orange. Moreover, we never experience reddish greens or bluish yellows. On the basis of such phenomenological observations, the nineteenth-century physiologist Ewald Hering was led to postulate two corresponding pairs of antagonistic processes in the visual system, a yellow–blue process and a red–green one. Now, contrary to what one might read in textbooks, there is no physiological evidence in the visual system for cells that secrete the sensations of yellowness and blueness or redness and greenness. The two subsystems found in the retina and visual pathway simply do not correspond to Hering's two axes. A light that varied between pure yellow and pure blue, for example, would strongly modulate both subsystems, whereas a variation between violet and lime yellow is what is needed to stimulate exclusively the ancient colour system (see Figure 9).

But do the two subsystems remain independent at a cortical level? Michael Webster and I have been tackling this question by measuring the loss of saturation that occurs for some hues if the eye is adapted to a repetitive variation in colour. A reason for mentioning these experiments is that they illustrate one of the analytical tools available to experimental psychologists. By adapting to a repetitive stimulus of a carefully designed kind, we can seek selectively to adapt a particular mechanism that we suppose may be within the nervous system. In the present case we adapt the eye to variation along a particular axis in colour space (Figure 9). By afterwards probing sensitivity in this and other directions, we can estimate how narrowly turned are the central channels that represent colour and how many there are.

In our experiments, the subject views a computer-controlled colour

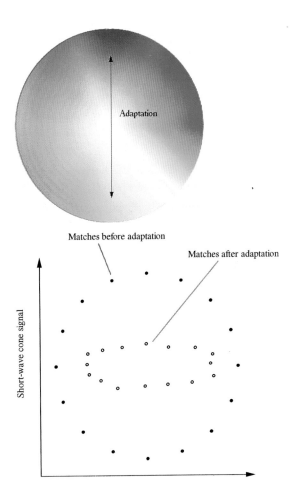

Figure 9 Probing central channels of colour vision by adaptation. The particular adaptation axis indicated above (from lime green to violet) is the one that would selectively adapt the ancient subsystem of colour vision. The subject was asked to adjust a reference colour falling on unadapted retina to match a test colour falling on adapted retina. In the graph below, coloured stimuli are plotted according to the signals they produce in the two subsystems, the vertical axis corresponding to the ancient subsystem and the horizontal axis to the newer one. The data points represent the appearances of colours before (●) and after (○) adaptation along a vertical axis: there is a selective loss of saturation along the same axis. The maximal loss of sensitivity is always along the adapting axis, whatever its direction (see text).

monitor. His or her task is to adjust the colour falling on an unadapted retinal area so as to match the appearance of a probe colour that falls on an adapted area. The adapting stimulus is an illuminated patch that varies rhythmically in colour along a particular axis of colour space, changing, for example, between a red and its complementary bluish green once per second. For all adapting axes, we ensure that the average stimulus over time is equivalent to a white light. The advantage of this kind of adapting stimulus is that, in the long term, there should be no differential adaptation of the retinal cones themselves; indeed we do not observe the common-or-garden complementary after-effects that one sees after simply staring at a coloured patch. Thus we are able to probe the neural representation of colour at a more central level.

In the graph of Figure 9, the outer solid circles represent the subject's match before adaptation. Suppose now we view an adapting field that varies between violet and yellow-green and thus exercises only the ancient subsystem. The open circles in the figure represent the match after adaptation. Colours that lie along the adapting axis then look very desaturated and plot closer to white, at the centre of the diagram, whereas colours on the orthogonal axis are relatively little altered in appearance. A similarly selective result is seen if the adaptation is along a horizontal axis from red to bluish green: now the greatest loss of saturation is found for stimuli along the horizontal axis. But if these axes were cardinal, if they corresponded to two classes of cortical neuron that between them represented all colours, then what should we expect if we adapted along a 45° axis in this space? If colour were represented only by the two cardinal axes, there should now be an equal loss of saturation for all colours. But in fact we again find that the saturation losses are largest along the axis of adaptation, and this is true for whatever direction in colour space we try.

This suggests that at the cortical level, there are neural channels tuned for many different directions in colour space. We do observe some enhanced selectivity for two axes, but these axes correspond to the ancient and modern subsystems identified physiologically in the visual pathway – and not to the putative red–green and yellow–blue processes of Hering. The special phenomenal status of the four pure hues is perhaps the chief unsolved mystery of colour science.

Colour constancy

I must deal finally with the relativity of our colour perception. The great geometer Gaspard Monge has attracted from his countrymen the supreme tribute of an eponymous Metro station, but what to me is a mark of his distinction is that he held high office, and retained his head,

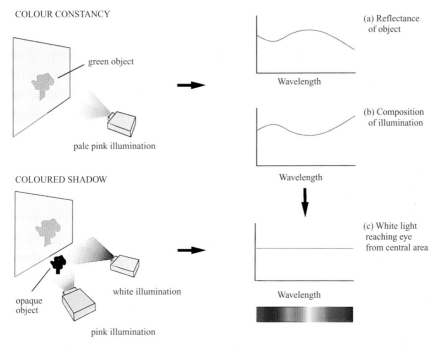

COLOUR CONSTANCY

green object

pale pink illumination

COLOURED SHADOW

opaque object

white illumination

pink illumination

(a) Reflectance of object

Wavelength

(b) Composition of illumination

Wavelength

(c) White light reaching eye from central area

Wavelength

Figure 10 A coloured shadow as a demonstration of colour constancy. At the bottom left a white screen is illuminated by light from two projectors, one giving pink light and one giving white light. An opaque card in the pink beam casts a shadow on the screen. This area therefore reflects only white light to the eye (graph *c*), but it appears a quite vivid, bluish green to the audience. Above is shown how a truly blue-green object would present the same stimulation to the eye if it were presented on a white ground in a pale pink illumination (top left). If we multiply the strength of the illumination at different wavelengths (graph *a*) by the proportion of light reflected from the object at each wavelength (graph *b*), then we derive (*c*) the relative intensity of different wavelengths reaching the eye from the central area. Graph (*c*) is identical in the case of 'colour constancy' and in the case of the 'coloured shadow'. It is also the case – and this is crucial – that the light reaching the eye from the surround is identical in the two cases.

in every administration from the *ancien régime* to the First Empire. On the eve of the Revolution, in May 1789, he gave to the Académie Royale des Sciences the most brilliant lecture that has ever been given on colour perception. We know, from an eyewitness account, the experiment with which he began: on the wall of a house that faced the windows of the Academy, he had fixed a sheet of red paper. He invited his fellow *académiciens* to look through a red glass and consider the colour of the red paper. The result was as counterintuitive in 1789 as it is today. We might expect that the red paper seen through red glass would have looked a vivid scarlet. In fact, it appeared whitish, and much less strongly coloured than when viewed with the naked eye. The experiment is today easy to repeat using a piece of red gelatine, and – as Monge himself remarked – the effect is most striking when the visual scene is a complex one, containing a variety of surfaces of different colours as well as the red target. A busy carpark serves well.

Monge realised that this was not an isolated illusion. He recognised that it shared an explanation with another phenomenon, which was already ancient in 1789, the phenomenon of coloured shadows. We can illustrate this effect by illuminating a screen with two projectors, one casting white light on the screen, the other, pink light – so that the surface appears homogeneously illuminated with very pale pink (Figure 10). If now we introduce a small opaque silhouette into the pink beam, so that it casts a shadow on the screen, then the shadow will appear a rather vivid green. If the demonstration is well set up, the green of the shadow will seem much more saturated than the pale pink of the rest of the screen. And yet – here is the mystery – the area of the shadow is physically illuminated only by white light.

Monge explained, with the frightening clarity that can be achieved only in the French language, that coloured shadows and the paradox of the red filter are not isolated illusions. Rather they reveal the operation of a mechanism that serves us well every instant of the day. This is the process of colour constancy, the process that allows us to perceive the permanent surface colours of objects independently of the colour of the illumination. A piece of white paper stubbornly looks white, whether we examine it in the yellowish light of indoor tungsten or in the bluish cast of northern daylight. What our visual system is built to recognise is permanent properties of objects, their spectral reflectances, their permanent tendency to reflect one wavelength more or less than another. What we are not built to recognise is the spectral flux, the balance of wavelengths in the light reaching our retina from an object on a particular occasion. For the latter, the balance of wavelengths reaching our eye from the object, depends both on the surface properties of the object and on the colour of the illumination.

But how does the visual system separate the spectral composition of the illuminant from the spectral reflection of the object? Monge supposed that our visual system estimates the illuminant from the average spectral flux in the surrounding field. In the demonstration of coloured shadows, our visual system very reasonably takes the illuminant to be a pale pink. The shadow is an object that reflects proportionately less red light than does the rest of the field. An object that behaves this way in pinkish illumination must be a green object.

We can use Figure 10 to elaborate the relationship between the illusion we call coloured shadows and the normal process we call colour constancy. In the case of the coloured shadow demonstration, the light reaching our eye from the shadowed area is white (as shown on the right) and the surround is a pale pink. In the equivalent case of colour constancy, we have an actually green object in pale pink illumination. If we multiply the spectral composition of the illumination by the spectral reflectance of the green object, we again have spectrally flat light coming from the object. The visual system is very intelligent, but it is not so intelligent as to distinguish between two projectors and one; these two stimulus arrays are therefore equivalent.

So there is no fixed relationship between the spectral composition of light and the hue that we perceive at a local point in the scene. This is the conclusion drawn with such prescient clarity by Gaspard Monge in 1789, when he wrote:

> So the judgements that we hold about the colours of objects seem not to depend uniquely on the absolute nature of the rays of light that paint the picture of the objects on the retina; our judgements can be changed by the surroundings, and it is probable that we are influenced more by the ratio of some of the properties of the light rays than by the properties themselves, considered in an absolute manner.

In other words, we judge colours by the company they keep.

Further reading

Cronly-Dillon, J. R. (ed.), *Vision and Visual Dysfunction*, vols. 2, 6 and 7, London: Macmillan, 1991.

Furumoto, L., 'Joining separate spheres – Christine Ladd-Franklin, woman-scientist (1847–1930)', *American Psychologist*, **47** (1992), 175–82.

Jacobs, G. H., 'The distribution and nature of colour vision among the mammals', *Biological Reviews*, **68** (1993), 413–71.

Jordan, G., and Mollon, J. D., 'A study of women heterozygous for colour deficiencies', *Vision Research*, **33** (1993), 1495–508.

Mollon, J. D., and Bowmaker, J. K., 'The spatial arrangement of cones in the primate fovea', *Nature*, **360** (1992), 677–9.

Monge, G., 'Mémoire sur quelques phénomènes de la vision', *Annales de Chimie*, **3** (1789), 131–47.

Nathans, J., 'The genes for color vision', *Scientific American*, **260**(2) (1989), 29–35.

Neitz, M., Neitz, J., and Jacobs, G. H., 'Spectral tuning of pigments underlying red–green color vision', *Science*, **252** (1991), 971–4.

Pokorny, J., Smith, V. C., Verriest, G., and Pinckers, A. J. L. G., *Congenital and Aquired Color Vision Defects*, New York: Grune & Stratton, 1979.

Webster, M. A., and Mollon, J. D., 'Changes in colour appearance following post-receptoral adaptation', *Nature*, **349** (1991), 235–8.

Winderickx, J., Lindsey, D. T., Sanocki, E., Teller, D. Y., Motulsky, A. G., and Deeb, S. S., 'Polymorphism in red photopigment underlies variation in colour matching', *Nature*, **356** (1992), 431–3.

6   Colour in Nature

Peter Parks

The problem with a subject as broad as 'colour' is that in a desperate effort to do justice to the topic, we nearly always fail to say anything of any substance. For this reason, forgive me if I confine this chapter to my experiences of 'colour in Nature'. Be prepared, therefore for a very biased account and one which centres particularly around aquatic biology.

Before we can appreciate the occurrence and function of colour, it is important to realise that, from start to finish, we are utterly dependent upon our individual perception in all cases and other people's perception in many cases. We are also dependent upon technology's perception. Here is an example to start the ball rolling. We are all aware that books on marine life invariably show us glorious aquamarine-coloured underwater scapes, populated by gaudily coloured fish and coral. The blue of under-sea shots is in reality usually decidedly grey and green. If the photograph was originated on Kodachrome and it is not tampered with in the print-ing process, it will appear as in Figure 1(a). If it was originated on Ektachrome (also made by Eastman Kodak), it will appear flatteringly bluer (Figure 1(b)). For this reason, most underwater photographers choose Ektachrome film stock, so that their pictures appear bluer!

Individual perception can be equally deceptive. I, for instance, see things yellower through my left eye than my right. My original partner in Oxford Scientific Films could not detect a difference in the colour of a green and a red traffic light and a research worker in the Zoology Department in Oxford used to wear a brilliant purple corduroy sports jacket. When I tested the researcher's colour vision with a set of Ishihara colour-blindness test charts, it became apparent that to him there was no difference between dark grey and brilliant purple! So beware before you think that we all see the same colour.

Figure 1 Underwater scenes, using: (*a*) Kodachrome; (*b*) Ektachrome.

There is also, in colour perception, the problem of tradition. If you go to the movies, you will nearly always see night scenes (which incidentally are usually shot during the day) depicted in deep blue hues. Moonlight particularly is portrayed blue. Have you ever tried to photograph the moon? It comes out remarkably brown! Some film makers have tried night scenes in sepia tones, but they nearly always arouse comment. Tradition says, nights are blue. They are not! Well, yours might be, mine are not!

However, it is now known that under night conditions we use the rod cells in our retinas rather than the cones, so appreciation of colour is impossible in the dark, since rods detect only light and dark. It is quite possible that people differ in their visual appreciation of light and dark, however, so we cannot be sure that some of us do not 'see' moonlight in shades of blue. In camera terms, it is a little bit like changing film stock and we have already seen how unreliable that can be. This shift of spectral sensitivity from cone to rod vision is known as the Purkinje shift.

It is also important to realise that Nature, too, is capable of biasing colour. If you only ever saw an animal on one occasion, at sunset, you would be very hard put to describe its colour accurately to someone wanting to identify it by day. Similarly, if you only catch a glimpse of a colourful bird out of the corner of your eye, you will never know its true colour. If a double-decker London bus passed through your peripheral vision you might possibly be able to detect its redness. The reason for this is that under daylight conditions the black-and-white seeing rods are saturated by excessive light stimulation, and, although the closely packed cones that exist in the central part of your retina can appreciate colour even in very small angular doses, the much more widely spaced peripheral ones cannot detect the colour of the small bird, but may be able to detect the colour of the large bus. Try this for yourself. Many people deny that they are unable to detect colour peripherally, but if you test them with objects with which they are unfamiliar, you'll be surprised how bad they are at this exercise.

We should also be aware that amongst animals some – even the proverbial bull – discern colours much less well than we do. It is now known that the eyes of mammals (other than primates) have only two types of colour sensitivity. They have blue-sensitive and yellow-sensitive cones and therefore probably see their world in the same way as a red–green colour-blind man. Other animals, such as insects, are capable of seeing wavelengths of light beyond the range of our vision. So bees see into the ultraviolet. Owls, however, may see into the infrared.

Colour through dyes and stains

Some colour in nature arises from external agents and before we look into generated colour, we should cover this first. Probably more often than appreciated, animals and plants are stained by the chemistry of their surroundings. Many mosses and sedges are stained rusty red, by ferrous chemicals. Sphagnum moss can take on a good deal of iodine from its surroundings. Flamingos fed on beetroot greatly improve the rose pink colour of their plumage; but though we might be tempted to think of this as a staining effect, it is not. It results from metabolic processes that deploy the colour compounds into the substance of the feather keratin.

Elephants in Tsavo National Park definitely exhibit the soil colour all over their leathery skins, a direct result of mud bathing and mud squirting by adults and youngsters alike. Baby grey seals are often born yellow. This results from the pups having literally fouled their amniotic fluid surrounds within the mother and prior to birth. Urine is the colouring agent. Some insects purposely stain themselves when they release defence fluids. Many caterpillars excrete a distasteful yellow fluid upon disturbance. This is relatively temporary, but it is an effective deterrent to predators. Strangest of all, some of the horned toads of North America stain their eyes by squirting blood from the lids. The function for this is far from clear.

Some of the pigmentation in Nature is laid down in the tissues of the animal or plant in a stable, fairly permanent form. It is all the more strange, therefore when we come across examples where it is not. The beautiful dark green and crimson coloration of the tropical Turacos, rather primitive birds, is said to be water soluble. It would appear, that the red pigment turacin is fast in pure water, but dissolves in water containing the lightest trace of alkali. The green pigment may be more water soluble. This conjures up an amusing vision of a tropical downpour cascading down limestone rocks and onto a Turaco!

While we are on the subject of stains and dyes, it is interesting to note that walnuts liberate a strong dye known as quasha and oak contains tanins, which if fumed with ammonia can render it dark green through to black. Bog oak, i.e. oak buried in bogs or river beds or even the ever-eroding North Sea, is jet black, as beautiful in colour and texture as any ebony. The old piles of London Bridge were as black as the ace of spades when it was demolished prior to shipment to the United States.

The pollen, and therefore the anthers, of spring crocuses is the source of saffron, that gorgeous rich yellow dye, so favoured by Thai monks. If you have ever brushed across an open flower with your sleeve, you will know just how effective a dye it is.

Figure 2 Aplysioid sea slug showing its purple dye.

Marine animals, too, produce dyes and inks. Squid, including the common cuttlefish, produce India ink, a jet black particle dye that is squirted out in a slightly viscid stream of mucus and therefore hangs in mid-water, allowing the perpetrator to escape behind the smoke screen. The beautiful purple dye produced in abundance by the aplysioid sea slugs is an effective deterrent to anything molesting this slow-moving mollus (Figure 2). The floating bubble-raft snail, *Janthina*, also secretes a jet of purple dye if molested. The dye that *Aplysia* produces was used by the Romans to create the vivid royal purple colour of togas belonging to dignitaries (see Lyons, Chapter 8).

Incidental colour

Many colours in nature could be described as incidental. There would seem to be no functional reason for their presence. On the face of it, many biologists would refute such a statement. They would maintain that all animal and plant colour fulfils a function. Let me then raise a query: What function does internal colour have? What possible purpose does colour have, if it can never be seen?

Surely, the colour of our internal organs is a product of the chemistry and metabolism within our tissues. The colour of a liver surely does not serve a purpose as such. I would accept that the red colour of blood not only lends colour but in addition a purpose to complexions. Blushing is an important social signal. I would also accept that the sight of blood sometimes prevents an aggressor pursuing an attack. A bloodied nose invariably stops a playground bully pressing home the attack. However, the presence of blood hardly dissuades a lion from further savage mauling of a human victim; in fact, probably just the opposite!

Although green sometimes indicates putrefaction to humans, I do not believe the green of garfish bones has a function. They may be poisonous to eat and the green colour could be a useful reminder to us, but that is not likely to benefit the filleted garfish!

No, I do believe that many colours in Nature are accidental, but I accept that nearly all external colours have a function. In some, like the intricacies of bird plumage or butterfly wings, colours may have arisen as products of metabolism, but now they function as tokens of individuality,

Figure 3 Incidental colouring in a tube-dwelling worm.

sexual development, camouflage, warning, mimicry, mood or courtship. In this context, there must be many colours present in animals that inhabit the abyssal depths that either function very differently from what we imagine, or are truly incidental colours (Figure 3). Why are acanthephyrid prawns so brilliantly red when there is no light to illuminate them? Perhaps by the glow of bioluminescent secretions from themselves or other neighbouring denizens of the deep, the red colour is inky black?

Colour chemistry

Most of us tend to think that colour is a type of compound that simply comes in a wide range of hues and tones. We look at a box of children's paints and think they are basically chalk, water and colour! Not so. Colour arises from a huge range of widely varying chemistries. The white in our paint box may have chalky origins, but the yellow might contain

(a)

Figure 4 (a) and (b) *Porpita*, showing its oceanic blue pigment.

chromium; the blue, potassium; the green, copper, and so on. If you mix poster paints with water, you will find the mixing properties alone are different. White and yellow rapidly sink in water. Green tends to float, as does blue.

So colour comes from a great spectrum of compounds and many arise for metabolic reasons just as much as for coloration. Therefore, the ways in which these disparate substances have been manipulated over the eons

(b)

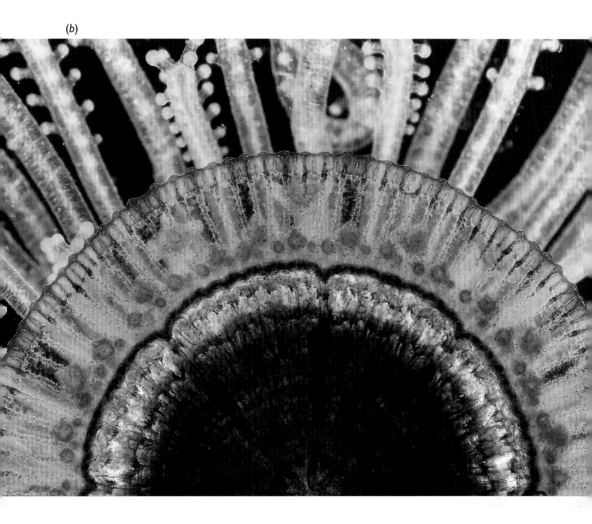

by natural selection (be it Darwinian and gradual, or Gouldian and explosive) is all that more remarkable, when we come to look at the incredible subtlety of the functions we now consider.

Some relatives of *Paramecium*, that familiar single-celled ciliate from stagnant freshwater, house chloroplast-rich algal cells. These render the species bright green and so vegetable-like, that only when they glide from cover are they clearly seen as animals. There is no doubt that throughout the Animal Kingdom, some colour hues are much more easily developed than others. In general, green, blue and purple pigmentation are rare. The Turaco bird is one of the few to actually metabolise and lay down a true green pigment. I am sure several readers will here take issue, and say that many birds and butterflies produce greens and blues in their coloration. True, but how many actually do it with pigment? Most produce those colours with structural colour and this I cover in some detail below.

There are some 'true blue' pigments produced in Nature, and many of the best of these are found in the sea. *Porpita*, a coelenterate, and many of its surface-drifting associates display gorgeous blues and purples (Figure 4). There is a brilliant blue copepod called *Pontellina*, which lives at the surface of the sea. This amazingly active but small crustacean, also lays claim to being the only copepod capable of leaping clear of the sea's surface. *Janthina*, the purple bubble-raft snail, mentioned above has a shell so richly coloured that collectors of shells prize its presence in their collections. *Porpita*'s cousin, *Velella*, commonly known as 'jack-sail-by-the-wind' sports a strong blue mantle and sky blue tentacles.

In the case of *Glaucus*, the silver-blue sea slug, which preys upon both *Velella* and *Porpita*, the coloration is nearly all structural. The silver sheen that adorns its ventral surface, is composed mainly of guanine in crystalline form. In *Physalia*, the Portuguese Man-of-War, each member of the colony, and therefore each tentacle type, is a different shade of blue, green, purple or pink, making a close-up view of this strange and ancient beast a feast of delicate and unusual colour (Figure 5).

While we are thinking 'aquatic' and 'blue', an interesting freshwater example occurs to me. *Paramecium*, that ciliated protistan inhabitant of ponds, particularly those into which farmers have dumped corn chaff, is colourless under normal illumination. If, however, it is looked at under high magnification with a suitably adjusted oil-immersion dark-field

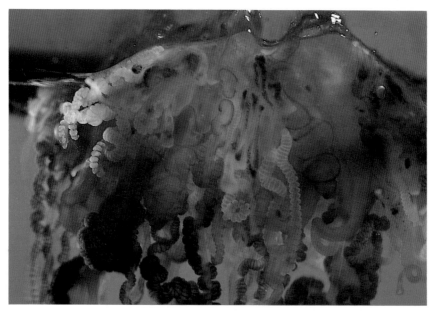

Figure 5 Physalia, exhibiting blues, purples and pinks.

condenser lens, then not only does the entire cell spring into vibrant colour and contrast, but its meganucleus appears a beautiful sky-blue colour. I have never heard a definitive explanation for this, but Ron Laskey, an old and eminent friend from Oxford, now Professor of Animal Embryology at Darwin College, Cambridge, suggested it could be a Tyndale-blue effect, which results from colloids in liquid suspension. The colour in this case would be a structural colour.

Structural colour

Structural colours in general have a familiar and common characteristic. They nearly all display a silk-like or metallic sheen: the blue speculum of a mallard duck; the blue flash of a Morpho butterfly wing; the purple hue to a snake that has just sloughed its skin; the blue-green of a kingfisher; the breathtaking colours of many birds of paradise, satin bowerbirds, hummingbirds, parrots and corvids. Innumerable insects, such as beetles, bugs and butterflies, have wings or wing covers that, if wetted with alcohol, will go dark brown or black. Upon drying, the original colour is restored. This is a simple and sure test for structural colour. The cause of the hue is invariably a series of closely spaced transparent scales, layers of chitin or keratin as in *Glaucus* (Figure 6).

When white light strikes and enters these thin transparent lamellae, it is split by refraction and diffraction into its spectral components, which then may further interfere with one another, according to angle. Thus, as we move past such a colour display, the hues change and shift. This is what characterises a metallic sheen or silk-like colour. The rainbow of colours that is seen when oil spreads on water is just such a structural colour and arises from the light repeatedly bouncing back and forth between the sandwiching interface of air/oil and oil/water. At each bounce, the blue light bends through an angle slightly different from that through which the red light bends. Bounce by bounce, this separation accumulates and the separate wavelengths interfere and yet more colours are created.

Sometimes on water, particularly at sea in the tropics, it is possible on a calm day to see natural oily slicks. These can be caused by blooms of diatoms, masses of fish eggs or even coral larvae. In this case, it is the individual transparent (or coloured in the case of coral) organisms that act like tiny prisms and droplets, thus splitting the light into its prismatic components.

Rain drops act in this way to produce a rainbow (see Longair, Chapter 3). A single rainbow is formed by the halo of drops in the sky which, with respect to the direction of the sun, subtend to the viewer's eye the appropriate angle so that light enters the drop, bounces once by total internal reflection, and then exits the drop. Depending on the precise angle, the viewer sees the drop and its neighbours as one of the colours of the rainbow. When a second, outer rainbow is visible, as when the sun is bright and the rain heavy, these drops have internally bounced the light twice.

Contrary to popular belief, it is possible to stand in one end of a rainbow. It is possible to do this also in the spray from a fountain or waterfall, but when the viewer is there, the colour effect is not very easy to see, and really the effect is in the form of part of a diverging tunnel, rather than a cylindrical limb, at one end of the rainbow. The critical feature is just where the drops are in relation to the sunlight. If literally half the sky was brilliant sunlight and the other half pouring rain, the effect could

Figure 6 Glaucus (eating *Velella*), showing structural blue.

be staggering. The height of the arch of the rainbow depends upon the elevation of the sun. A low sun makes a high rainbow; a high sun makes a low rainbow.

There is a small parasitic copepod that attacks salps and displays the most fabulous spectrum of structural coloration; its name is *Sapphirina*. The transparent laminated structure of its exoskeleton splits the light, and the intriguing thing is that every individual is different (Figure 7(*a*)).

Figure 7 Structural colour: (*a*) *Sapphirina*; (*b*) comb plates on a lobate ctenophore.

The beating comb plates of comb jellies, especially beroids and mnemiopsids, are also beautiful examples of pelagic marine structural colour. In the comb jellies, the spectral colours progress in unison up the eight comb rows in synchrony with the metachromal rhythm of the beating combs (Figure 7(b)). The flashes of colour are so bright as to be clearly visible from the deck of a vessel passing through the typical calm waters that sometimes swarm with these beautiful and archaic animals.

Warning colour

No less bright and gaudy are colour systems designed to give warning. These are termed aposomatic colours. Insects, reptiles and amphibians

Figure 8 Eyed hawk moth with aposomatic 'flash' eye spots.

are the experts in this behaviour, and the colour patterns and combinations some of their species display when threatened are just unbelievable.

The European tiger moths (garden, jersey, creamspot and scarlet) not only have warning colours on their forewings that are visible at rest, but also flash warning colours on their hindwings which are suddenly revealed if a bird or a vole should attack. The brilliant reds, yellows and oranges, offset by black and white, in these species are extremely effective.

Many moths have very cryptic forewings, and then surprise intruders by flashing bright and gaudy hindwings. In the case of the eyed hawk moth, the hindwing has a beautiful and fearsome eye on it which further enhances the effect (Figure 8). In the case of the emperor moth, the two eyes on the forewings suddenly become four eyes when the hindwings are also displayed. The bedstraw hawkmoth goes even further. It flashes, and then puts on a most convincing bee- or hornet-like dance, which, when you first see it, really startles you. These very few examples demonstrate just how closely associated with behaviour colour often is, and how colour strategy could seldom be effective without the appropriate behaviour. Even a superbly camouflaged insect or bird must learn to remain motionless in the presence of a predator if the crypsis is to work.

A nice example of warning colour in Amphibia is the fire-bellied toad, *Bombina*. This otherwise mottled green toad, if confronted by danger, be it snake or hedgehog, will arch itself backwards and curl its legs so that the brilliant red or orange palms to its feet and the side of its belly become clearly visible.

Colour mimicry

Pretending to be what you are not, as in the case of the bedstraw hawk-moth, leads conveniently on to the topic of mimicry. If one animal is going to mimic another, it obviously has to resemble it. This may take the form of shape or behaviour, but it also nearly always involves colour mimicry. Some examples of mimicry, especially in fish and butterflies, are extremely effective. Some are also very complicated. In the case of a hover fly called *Volucella bombylans*, different individuals mimic different types of bumble bee. Some mimic *Bombus lapidarius*, others *Bombus terrestris*, others *Bombus lucorum*, and even others *Bombus rupestris*. So it is possible to see several individual *Volucella bombylans*, each looking completely different from the next. Similar examples of polymorphic mimicry of a species that stings or is distasteful occur in moths that mimic wasps, hover flies that mimic honey bees, spiders that mimic ants, caterpillars that mimic snakes and, believe it or not, one South American plant hopper that looks uncannily like a young Cayman crocodile! Such mimicry is usually termed Batesian mimicry.

A second form of mimicry also occurs, and this is Müllerian mimicry, in which one distasteful species mimics another distasteful or even poisonous species. In general, if you are distasteful, the last thing you want is for a predator to learn on you. Much better if it learns on another species that you resemble. By the time it comes to you, with luck, it will have learnt not even to peck at you. Suck mimicry occurs in New World heliconiid butterflies that mimic very distasteful ithomiid butterflies: both are aposomatically arrayed in orange, white-and-black ornament – basic warning colours. It is interesting that humans also have adopted these colours for warnings on vehicles, waste tips and office signs.

Camouflage

Camouflage is a concept we all understand, from fighter aircraft to hen pheasants, and from centurion tanks to chameleons. The object wishing to be unnoticed has somehow to imitate or reproduce some of the colour and shape characteristic of its surroundings. There are many spectacular examples of camouflage and a few are worthy of mention. The Sargassum angler fish, *Histrio*, takes a lot of beating. Its outline, colour and movements blend with its floating habitat (Figure 9(a)). It even includes pattern and colour to mimic the chalky bryozoan and hydroid growths that colonise the weed. Flower mantids from the tropical Far East so closely resemble the orchids they inhabit that they just disappear from view. A beautiful transparent shrimp that lives on *Plumularia* hydroids in the Pacific has a gut inside its glassy frame that perfectly imitates the fibrous rachis of the hydroid. The shrimp always lies in line with the stem by day. Bottom-living scorpionid fish are among the most cryptic of marine animals. The notorious reef stone-fish *Synanceichtys* is no 'slouch' when it comes to looking like a stone. The fact that it can kill you if you stand on it is worth remembering, though!

Camouflage in some creatures is controllable, and in others it is changeable but less controllable, and in others it varies from individual to individual. Chameleons can change colour slowly to more-nearly resemble their immediate surroundings. To achieve this, pigment migrates within stellate pigment cells in response to visual signals. Cuttlefish and octopus can change colour very rapidly and can almost instantly effect camouflage to suit their backdrop. Flatfish can do the same in minutes. So, within a single population, individuals can look quite different, depending entirely upon what patch of sand they have last settled on.

The colour of some camouflaged animals is produced not only by the creature's own pigmentation but also by other organisms living in the tissues of the 'host' animal. These organisms are usually 'tolerated' or 'welcomed' because of the mutual benefit received. Sloths harbour green algal cells in their fur and therefore are even better mimics of their floral surroundings

(a)

Figure 9 Camouflage: (*a*) Sargassum angler fish.

(b)

(c)

Caption for Figure 9 (*cont.*)
(*b*) reef prawn; (*c*) hydroid-associating prawn.

than they otherwise would be. *Porpita*, that gorgeous chondrophoran co-elenterate that rides the tropical oceans, has symbiotic algal cells in its tissues that render its colour even more deeply oceanic (see Figure 4).

Amongst squid, the pigment is housed in tiny flattened bags, the walls of which are controlled by even smaller muscles. For this reason, squid can literally change colour in the 'blink of an eye'. The pigment migration in crustaceans, fish and chameleons, as well as the muscular control of pigment-containing cells in squid and octopus can be effective as mechanisms of camouflage, so long as the creature's visual perception of its immediate surroundings is the stimulus that activates the colour change. As might be expected, animals than can modify their colour to match their surrounding habitat may become camouflaged to a degree of awe-inspiring perfection.

Bioluminescence

Some animals give off bioluminescence. Some contain bacteria, which they stimulate chemically to glow, and some use filters to modify this light. Others produce their own alchemy of compounds, which when allowed to mix give off light. *Cypridina*, an ostracod crustacean, produces its own light chemistry. Some produce liquid secretions that glow upon deployment outside the animal's body. The deep-sea prawn, *Acanthephyra*, produces spectacular emissions of this sort. The purpose may be divertionary, since it tends to happen if the beasts are disturbed. Some bioluminescence has a sexual function. Fireflies certainly attract mates with their light displays, and several species of deep-sea squid and fish would appear to use bioluminescence in their courtship displays.

Transparency

Before bringing this account to a close, with the mechanism of colour control and colour change in Nature, it is worth emphasising that many animals choose to have no colour, or very little. Many marine animals, while in their larval pelagic planktonic phase, are so transparent as to be virtually invisible. Some of the more notable examples include *Leptocephalus* larvae of eels, *Phyllosoma* larvae of lobsters (Figure 10), and

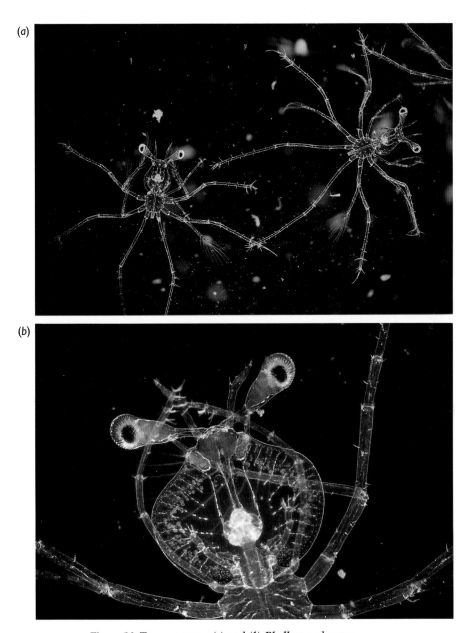

Figure 10 Transparency: (*a*) and (*b*) *Phyllosoma* larvae.

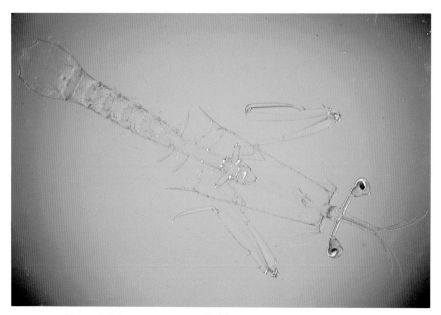

Figure 11 Transparency: squillid larva.

the wonderful larva of the mantis shrimp *Squilla* (Figure 11). Transparent adults include the deep-sea amphipod *Cystosoma* and the strange venus girdle comb jelly, *Cestus*, which may be over a metre long and flat as a belt.

Nearly all the transparent crustaceans, and even the fish larvae, do have a vestige of pigment. This is invariably associated with the eyes, in which retinal pigments always seem to exist, and also there is usually some colour associated with the gut. In *Cestus*, only the interference colour of the comb plates give any clue to the creature's presence. Not unlike transparency is the occurrence of mirroring. Fish particularly rely on mirrored scales to help them to appear invisible. The deep-sea hatchet fish *Argyropelecus* in this way uses mirrors on its flanks, and pigment on its top surface, to deceive its enemies.

Colour control and colour change

Last but not least I should say a little about colour control and colour change. In the Plant Kingdom there is a spectacular display of colour change each year in temperate and polar climes, when autumn leaves turn every shade of yellow, gold, orange and red, from their more typical lush green of summer.

The green of leaves is due mainly to the presence of chlorophyll *a* and *b*. The yellow and reds are due to xanthophyl and carotene. These pigments are there in the summer, too, only they are masked by the green pigments. In autumn the temperature change and the shortening of daylight periods causes the chlorophyll to break down, so revealing the yellow, golds, oranges and reds of the other pigments. New England colours in the United States are famous for their spectacular appearance. But the United Kingdom, too, can boast some spectacular shifts of colour amongst beeches and aspens. Those of us with Virginia creepers in our overfilled gardens will also be aware of a gaudy event put on by that plant, come the year's end.

In animals, colour change can relate to crypsis, mood change, sexual desire and passion, 'flight' or 'fight', luring of prey, and a host of behavioural quirks that are probably too subtle for us to understand. Such changes can be produced by more than one mechanism. Blood coursing more freely close beneath the skin can cause us to blush. Even rabbits have been shown to do the same thing! In crustaceans and fish, as I have already mentioned, signals from the eyes can cause pigment to migrate and therefore spread and contract within stellate pigment cells to effect colour change.

Conclusion

By way of conclusion, let me just emphasise that, within every aspect of the appreciation of colour, the observer must realise that what he or she sees may not be what others see, and therefore every one of us can rightfully consider that what we see of colour is something very private. Above all though, I am sure it is there to be enjoyed, and my heart goes out to all our fellow beings who, for reasons of mishap, disease or accident, are unable to share the pleasure that the rest of us experience. We should therefore so much the more appreciate our gift of seeing 'colour in Nature'.

Further reading

Buchsbaum, R., and Milne, L. J., *Living Invertebrates of the World*. Hamish Hamilton, 1960

Encyclopaedia Brittanica, 'Colour'.

Ford, E. B., *Moths*, 3rd edn, London: Collins, 1955.

Hardy, A. C. *The Open Sea–The World of Plankton*, 2nd edn, London: Collins, 1970.

Herring, P. J., and Clarke, M. R., *Deep Oceans*, London: Arthur Barker, 1971.

Meeuse, B., and Morris, S., *The Sex Life of Flowers*, London: Faber & Faber, 1984.

Parks, P., *The World You Never See: Underwater Life*, London: Hamlyn, 1976.

7   Colour and Culture

John Gage

A series of discussions on colour from different perspectives soon lands us in serious conceptual difficulties. I want to start by proposing that most of the authors in this Darwin series are not really dealing with 'colour' at all: they are concerned with radiant stimuli in light or with the physiological processing of these stimuli by the eye, whereas 'colour', properly speaking, does not come into the picture until rather later, in the mind which apprehends it. Research into a brain dysfunction called 'colour anomia' by J. and S. Oxbury and N. Humphrey, and more recently by A. Damasio, suggests a sharp distinction between the sensation of 'colour' and its identification. Patients with normal colour vision who were able to perform purely verbal tasks with colour names, such as naming the colour of a named object like a banana, were unable to name correctly the colours that they saw: blue, for example, was called 'red'. These patients, like all of us, used verbal language as the customary tool of communication; if 'colour' is intimately bound up with language – if it is a system of arbitrary signs – it must also be a function of culture and have its own history. And yet this linkage with verbal expression is highly problematic.

Colour-usage and colour-systems

A few years ago Umberto Eco published an essay under the title, 'How culture conditions the colors we see', but he was unable to live up to the promise of this ambitious formulation because of the very imprecision of his term 'culture'. As a semiotician, Eco was embarassed in his discussion of colour by the almost complete absence of intelligible codes of colour-meaning within a given culture. Does 'culture' imply knowledge and

Figure 1 After-image stimulus. © DACS 1995.

embody rational investigation, or may it run counter to them? Is it, in effect, largely a matter of assumption and prejudice? Who are the agents and guardians of 'culture'? Colour promises to throw some light on this problem because, in the Western societies that provide me with my material, colour-usage has long co-existed with more or less sophisticated theories of colour that are relatively well known. Several of these high-level theories have recently been discussed by Martin Kemp in *The Science of Art*, but this low-level colour-usage does not encourage a belief in the cultural coherence of codifiable systems of thought. Let me illustrate this with a few homely examples.

If you look intently at the red disc in Figure 1 for a minute or so, and then relax your eye muscles while looking at a piece of white paper (in each case fixing on the centres of the field of vision), most of you will see a colour that you will probably be inclined to call 'blue-green', a colour close to the one which Matthias Grünewald represented encircling Christ's red-orange halo in the *Resurrection* scene in his great Isenheim Altarpiece of the early sixteenth century, now in the museum at Colmar in Alsace. Grünewald, who may have been a technologist as well as a painter, had doubtless experienced this colour, as we all do, as the negative after-image of a fiery red light. When, in the late eighteenth century, the phenomenon of negative after-images began to be investigated systematically, notably by Charles Darwin's father, Robert Waring Darwin, the 'complement' to red was also usually described as blue-green, as it had been about a century earlier in Newton's experiments with the colours of thin plates ('Newton's Rings': *Opticks*, Bk ii, Pt i, Obs. 9). But after 1800, the notion that there are three 'primary' colours of light (red, blue and yellow) and that the eye, fatigued by the strong sensation of one of these colours, 'demanded' the product of the remaining two in order to restore its balance, was allied to an interest in symmetrical, usually circular, colour-systems. It became increasingly common to describe, and even to represent, the complement of red as simply green (see Figure 2), a mixture of equal parts of blue and yellow. Green is still commonly identified as the complement of red, even in perceptually oriented handbooks of colour such as Josef Albers' *Interaction of Color* (1963); and this persistent idea suggests a powerful cultural conditioning of the sort Umberto Eco was concerned to expose.

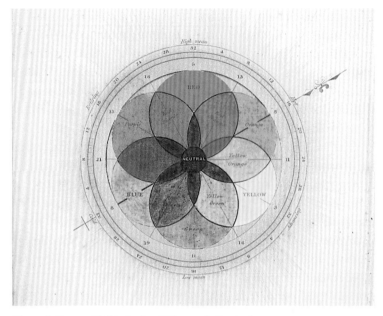

Figure 2 George Field, *Scale of Chromatic Equivalents.*

Figure 2, George Field's colour-circle of 1841, however, also introduces us to experiences of colour where culture seems to have worked in precisely the opposite way, where perceptions appear to take precedence over ideas. Field's polar contrasts 'hot' and 'cold' may here be making their first appearance in a colour-system, although they had been common enough in English painterly discussions for at least a century. But they are contrasts which are still widely endorsed in the characterisation of colour. Colours seem 'warm' or 'cool' only metaphorically, of course, but the radiation of which they are the visible symptom is radiant *energy*, and we have known ever since the introduction of gas heating over a century ago that it must be interpreted in the opposite sense to this metaphorical usage. The short-wave, high frequency energy of the blue-violet end of the spectrum signals the greatest capacity to heat, and the long-wave, low frequency red end, the least. Yet even in the modern world gas companies continue to show the warming effect of red-orange flames where domestic comfort in the living-room is in question, while they take a much more functional attitude to ovens, which are shown correctly with the heating flames as blue. Laboratory tests in Europe and the United States from the 1920s until the present day have shown that the psychological interpretation of colour-temperature has been far from unambiguous, but I imagine that most people will continue to think of yellows, oranges and reds as at the 'warm' end of the spectrum, and blues and greens at the 'cool'.

Colour universals

In recent years there has been a revival of interest in the idea of a universal or 'basic' experience of colour, which is seen to have given rise to these interpretations that jibe so much against what we may be assumed to know. Responses to colour, it is argued, go back to archetypal human experiences of black night, white bone, red blood and so on. Thus A. Wierzbicka proposed in 1990 that:

> yellow is thought of as 'warm', because it is associated with the sun, whereas *red* is thought of as 'warm' because it is associated with fire. It seems plausible, therefore, that although people do not necessarily think of the color of fire as red, nonetheless they do associate red color with fire. Similarly, they do not necessarily think of the color of the sun as yellow, and yet they do think of yellow, on some level of consciousness or subconsciousness, as of a 'sunny colour' . . .

The problem with this approach, which is rather widespread among anthropologists, is that the stable referent has usually been more interesting and important than the colour. The colour of familiar phenomena in Nature has, indeed, often been a matter of puzzlement and debate. As early as the first or second century A.D. the Greek astronomer Cleomedes pointed out the variety of colours in the sun, now whitish or pallid, now red like ochre or blood, now a golden or even a greenish yellow, and only sometimes the colour of fire (*Caelestia* II, i). And in the fourth century St Augustine of Hippo even made the various colours of the sea green or purple or blue – all of which may still be readily seen in the Mediterranean – one of the touchstones of variety in nature (*The City of God*, XXII, xxiv).

Recognising colours: the spectrum in the rainbow and the natural world

There are good reasons for thinking that a precise recognition of all the colours we are capable of discriminating has often been a matter of indifference. Languages have never been used for labelling more than a tiny fraction of the millions of colour-sensations which most of us are perfectly well equipped to enjoy and, we might have supposed, to name. A glance at a standard modern handbook of colour names such as A. Maerz and M. R. Paul's *Dictionary of Color*, which lists and represents mainly English-language tradenames, will show that, although most of us are perfectly capable of discriminating among an extensive continuum of

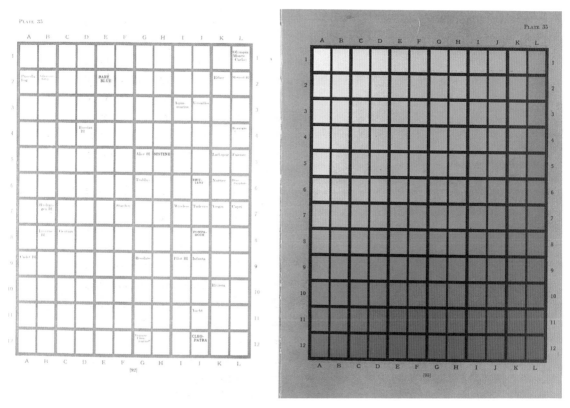

Figure 3 A. Maerz and M. R. Paul, *Blue.*

colour-nuances very few of these nuances have been named (Figure 3); and modern colour-systems, following the lead of James Clerk Maxwell in the 1860s, have usually resorted to numbers in order to distinguish perceptible differences of hue or value (lightness or darkness) in what has turned out to be a far from symmetrical colour-space.

Probably the most widely recognised of these colour continuums in the ancient and modern worlds has been the spectrum of light as manifested in the rainbow, It was the optics of the seventeenth century, notably the work of Sir Isaac Newton, that made the spectrum into the standard of 'colour'; and it is striking that in the eighteenth century even a natural philosopher such as the Viennese entomologist Ignaz Schiffermüller, whose concern with colour was primarily as a means of identifying butterflies, should have used the spectrum, as an indoor experiment and, as an outdoor phenomenon of nature, as the paradigmatic manifestation of colour in the illustration to his *Essay on a System of Colours* (*Versuch eines Farbensystems*) of 1772 (Figure 4). Yet the number and even the order of colours in the rainbow has always been a matter of dispute. Although Newton's isolation of seven spectral colours had been anticipated by Dante in *The Divine Comedy* (*Purgatorio* xxix, 77–8), and illustrated very plausibly in the scene of Noah's Flood in a fifteenth-century Norman Book of Hours now in the Bodleian Library in Oxford, Newton repeatedly changed his mind during the course of his career and, as Malcolm Longair points out (see Chapter 3, p. 72), opted for the seven-colour version only because he was anxious to sustain an analogy with the musical octave. The patriotic English Romantic John Constable, who was famous for his sharpness of observation, seems nevertheless, in his frequent depictions of the rainbow, to have been content with red, white and blue; and in the modern world of commercial design I have found examples with five, six or seven colours, and a variety of sequences. It is chiefly the imperceptible transition from one band of colour to the next which has led to these ambiguities, and it is not surprising that we sometimes have to resort to mnemonics to remember the order.

Although Constable, a painter who showed an unusual interest in meteorology, correctly recorded the reversal in the sequence of colours in the secondary rainbow, which is sometimes observed outside the first, this has not always been respected. One of the most unexpected lapses was in

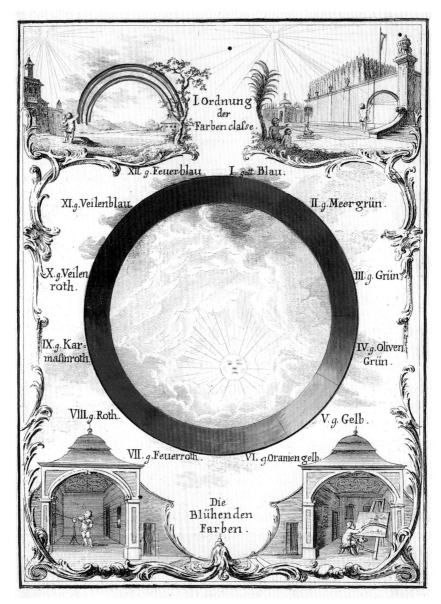

Figure 4 Hand-coloured illustration from Ignaz Schiffermüller, *Versuch eines Farbensystems.*

John Everett Millais's *Blind Girl* (1856), now in Birmingham City Art Gallery, where the Pre-Raphaelite precision of the landscape setting, painted at Winchelsea in Sussex, is quite remarkable, but the colours of the secondary bow were not reversed until the mistake was pointed out by a friend of the painter's, and corrected – for a supplementary fee. Even in the case of a single bow the order of colours, running from red at the top to violet inside the arc, has sometimes escaped the attention of artists. Some readers may recall the broad upside down bow in the *Pastoral Symphony* section of Walt Disney's *Fantasia* of 1940.

What the history of the spectrum suggests is that there are real difficulties in perceiving and identifying colours in complex arrays, especially when their edges are undefined, and that the relative poverty of colour-vocabularies reflects these difficulties and encourages representations to be far more concerned with ideas about colours than with colour-

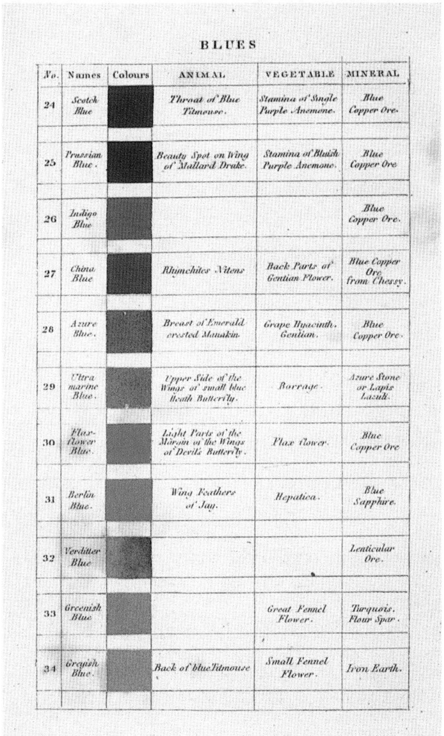

BLUES

No.	Names	Colours	ANIMAL	VEGETABLE	MINERAL
24	Scotch Blue		Throat of Blue Titmouse	Stamina of Single Purple Anemone	Blue Copper Ore
25	Prussian Blue		Beauty Spot on Wing of Mallard Drake	Stamina of Bluish Purple Anemone	Blue Copper Ore
26	Indigo Blue				Blue Copper Ore
27	China Blue		Rhynchites Nitens	Back Parts of Gentian Flower	Blue Copper Ore from Chessy
28	Azure Blue		Breast of Emerald crested Manakin	Grape Hyacinth Gentian	Blue Copper Ore
29	Ultramarine Blue		Upper Side of the Wings of small blue Heath Butterfly	Borrage	Azure Stone or Lapis Lazuli
30	Flax-flower Blue		Light Parts of the Margin of the Wings of Devil's Butterfly	Flax flower	Blue Copper Ore
31	Berlin Blue		Wing Feathers of Jay	Hepatica	Blue Sapphire
32	Verditter Blue				Lenticular Ore
33	Greenish Blue			Great Fennel Flower	Turquois, Flour Spar
34	Greyish Blue		Back of blue Titmouse	Small Fennel Flower	Iron Earth

Figure 5 P. Syme, *Blues.*

perceptions themselves. The devising of colour-systems allowing colours to be set out in a logical sequence which articulates relationships between them, scarcely pre-dates the seventeenth century; and if the spectrum of white light, especially after Newton rolled it into a circle in his *Opticks* of 1704, was embraced as the most coherent of these systems, it not only remained the conceptual problem to which I have alluded but was still impossible to translate into terms of surface colours, because of the impurities in the available pigments and dyes. Aristotle had already argued that the pure colours of the rainbow were impossible to represent in painting, and well into the nineteenth century colour-atlases for the use of naturalists might avoid the spectral colours and base their standards of hue and value on a range of natural objects. Patrick Syme's 1821 adaptation of A. G. Werner's *Nomenclature of Colours*, a copy of which accompanied Darwin on the Beagle, is a prime example of this (Figure 5), as it is also of regional variations in colour-categorisation, even within the scientific community at this date. Werner, a late-eighteenth-century German mineralogist, had subsumed purple under blue and orange under yellow, but Syme, a Scottish flower-painter, argued that purple and orange were as entitled to be considered independent colours as were green and brown (two colours which, incidentally, have retained their *psychological* independence until our own day). It is perhaps worth noting that Darwin does not seem to have used the nuanced and hence rather precise Symian/Wernerian terminology in his taxonomical reports for long after the *Beagle* expedition of the 1830s. It was probably far too cumbersome for regular use.

Colour specification for scientists has now become exclusively mathematical, but it is, of course, only the stimuli that can be quantified, not the 'colours'; and as recently as the 1940s an English pioneer of non-representational painting, Winifred Nicholson, devised a spectrum of hues and values entirely related to natural objects (Figure 6). Her scale underlines the fact that surface colours possess several characteristics apart from the hue, value and saturation (chroma), which have usually been held to define the parameters of colour as perceived. One of these characteristics is texture, and especially among Russian artists and critics around the time of the First World War, texture (*faktura*) came to be recognised as a specific aesthetic category. The radically non-

clay	mud	dust	earth	shadow	slate	lead
terracotta	dun	putty	khaki	mist	pewter	prune
brick	fawn	beige	faded oak leaf	sea grey	steel	mulberry
roan	bistre	hay	sage	air force blue	blue grey	vieux rose
rust	ochre	straw	willow	fell blue	knife blue	musk rose
coral	sand	amber	crab apple	turquoise	royal	wine
ruby	flame	topaz	emerald	azure	sapphire	amethyst
RED	**ORANGE**	**YELLOW**	**GREEN**	**BLUE**	**INDIGO**	**VIOLET**
sugar pink	alabaster	sulphur	duck's egg	baby ribbon blue	ice blue	pale lilac
scarlet	apricot	lemon	pea green	sky	french blue	lavender
vermilion	fire	canary	grass green	forget-me-not	hyacinth	heliotrope
tomato	fox	brass	cabbage	larkspur	ultramarine	purple
dragon's blood	copper	daffodil	forest green	lapis-lazuli	electric blue	maroon
mahogany	tobacco	mustard	laurel	horizon	midnight	damson
RAVEN	BLACK COFFEE	TIGER SKIN	BLACK VELVET	ZENITH	PITCH	CHOCOLATE

Figure 6 Colour scale by Winifred Dacre (Nicholson).

representational works painted by Kasimir Malevich under the banner of 'Suprematism', for example, depended, in their articulation of several whites, partly on very subtle textural variation. One of the great masters of texture appealed to by the Russians was the Monet of the late *Rouen*

Cathedral series, in which an almost relief-like handling of surface texture was one of the most significant of his painterly tools.

Basic colour-terms

But such a wide-ranging understanding of the phenomenology of 'colour', although it has a substantial history going back to Classical Antiquity, and has been explored extensively by twentieth-century psychologists such as David Katz, runs counter to the usual modern conception of the phenomenon, which, at least since Newton, has focused almost exclusively on the characteristic of hue; that is, on spectral location. The widespread interest aroused not only among ethnologists and linguists (see Lyons, Chapter 8) but also among semioticians and even physiologists by Brent Berlin and Paul Kay's *Basic Color Terms* (1969), depends very largely on a remarkable convergence of experimental findings between these two eth-nologists (who argued for the universal recognition of eleven 'basic' colour-categories whose foci were located by their subjects on a spectrally arranged chart of Munsell colour-chips) and some modern students of the mechanisms of colour vision, who have identified a reduced set of colour receptors in the retina, arranged to process pairs of 'complementary' or 'opponent' stimuli – red-green, blue-yellow and light-dark. Berlin and Kay identified their eleven 'basic' terms in nearly one hundred widely scat-tered languages, and even the far larger sample in the World Color Survey, since initiated by them, has hardly modified the structure of their underlying scheme. As the distinguished anthropologist Marshall Sahlins has commented: 'it is difficult to escape the conclusion that the basic color-categories are natural categories'. Sahlins was unhappy with this inference, since he supported a subtle version of the cultural relativism Berlin and Kay's research was proposing to combat; but he might have escaped from it rather easily had he taken on board the curious consider-ation that to the physiologists' six categories (listed above) even Berlin and Kay's 'basic' set adds five others, including grey, pink and brown. Their definition of 'basic' has certainly come under fire from T. D. Crawford, and more comprehensively from J. van Brakel and B. Saunders; an examination of the history of the notion of 'basic' colour sets – often assimilated to the concept of the four elements, earth, air, fire and water, in Classical Antiquity and the Middle Ages – shows that they shared almost no other common feature than the desire to reduce the myriad of colour-sensations to a simple and orderly scheme. As a leading modern student of the relationship between psychology and aesthetics, Rudolf Arnheim, puts it:

> Neither man nor nature could afford to use a mechanism that would provide a special kind of receptor or generator for each color shade.

'Basic' sets of 'simple' or 'primary' colours are thus a great gift to structuralists, but they offer little comfort to those of us who are concerned to interpret the use of colour in concrete situations.

Marshall Sahlins shares with other modern thinkers, notably Ludwig Wittgenstein, a belief that the assumptions embodied in 'ordinary' colour-language reflect the logic of modern colour-order systems of the Munsell type. Thus he writes:

> Blue is always different from yellow, for example: depressed ('the blues'), where yellow is gay, loyal ('true-blue'), where yellow is cowardly, and the like. Blue has a similar meaning to yellow about once in a blue moon.

It is true that many Western cultures have taken on board these modern systems, with their emphasis on contrasting hues; but there are instances in the Western Middle Ages, as well as in several non-Western languages spoken today, where the same term was used to cover both blue and yellow, including the Old French word *bloi*, the ancestor of our English 'blue'. Similarly, the other pair of 'opponent hues' in modern perceptual theory, red–green, was also covered by a single term in the Middle Ages: *sinople*, a red in Old French poetic usage, but green in the specialised vocabulary of heraldic blazon, which was also French. Even in our own times, Wittgenstein's nonsense term 'reddish green' (*Remarks on Colour*, I, 9–14) has been perfectly acceptable in some languages (for example one spoken in parts of Japan), and even in Germany in the 1920s in the context of Paul Klee's design teaching at the Bauhaus. The anthropologist R. E. MacLaury has recently drawn attention to instances of non-European languages where white has been assimilated to black; it is clear that in some cultures which have had a profound effect on Western colour ideas, notably ancient Judaism, the semantic polarity of positive and negative, which has usually been regarded as axiomatic for this pair (white = positive; black = negative), does not apply. An important tradition of Christian mysticism deriving from Pseudo-Dionysius in the early Middle Ages proposed that darkness was, indeed, the very seat of God. These apparent anomalies have been noticed only recently, and need much more investigation; but they suggest that in the case of Western societies as well as in non-Western ones, colour-usage cannot always be understood in terms of colour science.

A disdain for colour

One of my favourite episodes in recent research into colour language is the arrival in 1971 of the Danish anthropologists R. Kuschel and T. Monberg, armed with their sets of Munsell colours, on Bellona Island in Polynesia, only to be told by a native, 'we don't talk much about colour here'. In the event their report seems to me to be one of the best modern investigations of colour-usage within a given culture, but it makes clear that colour as hue is not everybody's interest, and in many contexts we can, of course, do perfectly well without it. The black-and-white photography, which in Charles Darwin's day seemed to offer a new touchstone for the visual representation of the real world, was only the latest phase in the history of monochrome reproduction, which goes back to Classical times. Darwin himself who, as an undergraduate, had frequented the Old Master print collection of the Fitzwilliam Museum in Cambridge, was content to use black-and-white engraving even to illustrate his discussions of the highly coloured plumage of exotic birds, for example in his *Descent of Man* (1871). At least since the Renaissance, sculpture in the Western tradition has also been largely uncoloured, partly because ancient Greek sculpture was thought, quite wrongly, to have cultivated this monochrome convention. Modern studies of colour vision have tended to reinforce the fundamental role of the rods in the human retina, which process variations in brightness or value; and it is well-known that colour-blindness may pass unnoticed for many years because it is scarcely a functional disability.

In some European and Oriental cultures, moreover, a disdain for colour has been seen as a mark of refinement and distinction. The taste for black in clothing, for example, was a prerogative of wealth and nobility in the Renaissance, but in succeeding centuries it spread in Europe to all levels of society, and black still forms part of our dress-code for many occasions. On the other hand, when Vincent van Gogh made painted versions of some Japanese prints in his collection, and substituted strident complementaries for the more subtle and reticent colour-harmonies of his models, it may seem to us to be no more than a case of ignorance and vulgarity. Yet it was one of the important achievements of the experimental psychology of van Gogh's time to have shown that a love of strong, saturated 'primary' colours was not the preserve of primitives or of children, but was also common among educated European adults, and this was a line of research which went hand in hand with the development of a new range of bright synthetic pigments and dyes. It was these psychological as well as technological developments that lay behind what has always been recognised as the enormously expanded interest in highly contrasting hues that marks the visual expression of twentieth-century Western culture, and which has sometimes been characterised, rather misleadingly, as the emancipation of colour in the modern world.

Colour psychology: chromotherapy and the Lüscher Test

The attempt to yoke the structures of colour-language to the mechanisms of colour vision, although widespread in recent years, is still a rather specialised academic pursuit, but another modern development from late-nineteenth-century psychology has had far wider ambitions. The belief that exposure to variously coloured lights could have a direct and variable effect on human bodily functions (it had already been studied in relation to plant growth, by Darwin among others) was perhaps first proposed in the quantified experiments by the French psychologist Charles Féré in the 1880s (*Sensation et Mouvement*, 1887, 41–6). Féré found that red light had the most stimulating effect and violet the most calming; but for the student of *visual* culture his ideas can only have a limited application, since he treated coloured lights simply as variable vibrations of radiant energy that could be sensed by the skin even with the eyes closed. This was the sort of research lying behind the development of chromotherapy, a practice which seems to have had its greatest vogue in

Europe around the turn of the century, but which is still in the repertory of alternative medicine. As a recent review by the physiologist P. K. Kaiser indicated, chromotherapy proved highly resistant to systematic analysis; but another branch of colour psychology, which proposes isolating non-associative scales of colour-preference, based on laboratory testing, has been more widely acceptable, perhaps because it is promoted and used by powerful marketing organisations for commercial purposes.

The most familiar of these scales is perhaps the test devised in the 1940s by the Swiss psychologist Max Lüscher, and which, according to his organisation, is now used widely in ethnographical research, in medical diagnosis and therapy, in gerontology, marriage guidance and personnel selection. The Full Test uses 73 colour patches, but a short and handy version includes only eight samples: dark blue, blue-green (also called 'green'), orange-red, bright yellow, brown, black and grey. The subject is asked to arrange the coloured cards in a descending order of preference, and an analysis of this order over a number of test runs allows the psychologist to interpret the subject's character. This interpretation is predicated on the meanings attributed to the colours. Thus blue, which Lüscher, like most modern psychologists, has identified as the most widely preferred colour among Europeans, is held to be concentric, passive, sensitive, perceptive, unifying and so on, and thus to express tranquillity, tenderness and 'love and affection'. Orange-red, however, is eccentric, active, offensive, aggressive, autonomous and competitive, and hence expressive of desire, domination and sexuality. The section of the public to whom the Lüscher Test is chiefly directed may be inferred from the interpretation it gives to an ordering which puts blue at the beginning and red towards the end of the sequence:

> Psychologically, this combination of rejected red and compensatory blue is often seen in those suffering from the frustrations and anxieties of the business world and in executives heading for heart disease ... Presidents, vice-presidents and others with this combination need a vacation, a medical check-up and an opportunity to re-assemble their physical resources.

What the English version of the *Lüscher Short Test*, edited by Ian Scott, does not reveal, is that the conceptual framework for these interpretations was found by Lüscher largely in German Romantic theory, in

Goethe's *Theory of Colours* (1810) and in the Neo-Romanticism of the early-twentieth-century non-representational painter Wassily Kandinsky, whose treatise *On the Spiritual in Art*, published in Germany in 1911 and in England and Russia a few years later, includes perhaps the most important body of colour ideas for modern artists. At one point in the first German edition of his book, published on the anniversary of Goethe's birth in 1949, Lüscher even introduced the ancient and mediaeval notion of the correspondence of the four humours and the four elements, with one of the sets of colours attributed to them. Goethe and his fellow poet Friedrich Schiller had been working along similar lines at the close of the eighteenth century.

It is thus no surprise that the Lüscher Test has come in for a good deal of criticism from psychologists for its dogmatic tone, and in particular for its failure to provide a uniform standard in the colour samples of its various editions. For the historian of culture its chief weakness is that it gives no consideration to the crucial question of whether the psychological response to colours is chiefly to their names, and hence to a general concept of each of them, rather than to their specific appearance. Recent work with animals and with infants might suggest the latter, were it not that the effects of exposure to colours have hitherto been so limited. But if language is crucial, the problems inherent in colour-vocabularies outlined above must be brought into play. Nevertheless the Lüscher system certainly rests on what seems to be a universal urge to attribute affective characters to colours, and it must be taken, at the very least, as an important modern manifestation of that urge.

Conclusions

In this brief discussion I have been able to touch on only a few of the issues raised by the theme of colour and culture. I have laid most emphasis on the instability of colour-perceptions because it should give pause to those many ethnographers and semioticians who have been tempted to speak confidently of colour-meanings and preferences in many cultures. I have hardly mentioned perhaps the most important issue of all: the definition of culture itself. Which sector of a given society is in question? Which gender, which age group, which class, which profession? In the case of aesthetic preferences we have seen both a liking for black spreading from aristocratic to general usage, and a taste for bright colours from ill-educated to educated groups. R. E. MacLaury has recently argued for an emphasis on brightness or value in colour-language as reflecting a belief in unity, and an emphasis on hue as indicating a belief in perceptual diversity. Yet Bridget Riley shows (see Chapter 2) that, at least for the specialised class of painters, hue itself has often been a tool for unification. Similarly the widespread perception that women are more discriminating than men in their use of colour may be linked to the relative rarity of colour deficiencies in female vision (see Mollon, Chapter 5).

Most of my examples, from Grünewald to Winifred Nicholson and from Cleomedes to Lüscher, have come from what used to be called 'high' culture, which at least has the advantage that it is more accessible to us than popular culture in the historical record. But I cannot resist ending with an example of the ways in which modern consumerism has appropriated the allure of this 'high' culture for the purposes of mass-marketing. Some years ago a British household paint manufacturer produced a range of emulsions and gloss colours which were launched under the names of a number of European Old Masters. Anyone familiar with the history of painting might well be bemused by the faded gentility of 'Turner' (pale violet) or 'El Greco' (pale blue), and equally perplexed by the close proximity of 'Chardin' to 'Vermeer' (both pale grey-greens). If by now you are thoroughly confused at how little the languages of colour relate to its perception, you may at least take heart that this manufacturer was prepared to supply a handful of these 'colours', from 'Leonardo' to 'Manet', in black and white versions as well. It may also be a welcome sign that this range of paints did not enthuse the do-it-yourself public, and the names of great artists were apparently soon replaced by numbers.

Further reading

Albers, J., *Interaction of Color*, New Haven, CN: Yale University Press, 1963 (paperback abridgement, 1971).

Arnheim, R., *Art and Visual Perception: the New Version*, Berkley, Los Angeles: University of California Press, 1974.

Berlin, B., and Kay, P., *Basic Color Terms*, Berkley: University of California Press, 1969 (reprinted 1991).

Boyer, C. B., *The Rainbow: from Myth to Mathematics*, New York, London: Thomas Yoseloff, 1959.

Conklin, H. C., 'Color categorization', *American Anthropologist*, **75** (1973), 931–42.

Crawford, T. D., 'Defining "basic color term" ', *Anthropological Linguistics*, **24** (1982), 338–43.

Damasio, A., 'Disorders of complex visual processing: Agnosiasis, Achromatopsia, Balint's syndrome and related difficulties of orientation and construction.' In *Principles of Behavioral Neurology*, ed. M.-M. Mesulam, pp. 259–88, Philadelphia: Paris, 1985.

Eco, U., 'How culture conditions the colours we see.' In *On Signs*, ed. M. Blonsky, pp. 157–75, Oxford: Blackwell, 1985.

Eysenck, H., 'A critical and experimental study of colour preferences', *American Journal of Psychology*, **54** (1941), 385–94.

Gage, J., 'Color in western art: an issue?', *Art Bulletin*, **72** (1990), 518–41.

Gage, J., *Colour and Culture: Practice and Meaning from Antiquity to Abstraction*, London: Thames and Hudson, 1993.

Izutsu, T., 'The elimination of colour in Far Eastern art and philosophy.' In *Color Symbolism: Six Excerpts from the Eranos Yearbook 1972*, ed. S. Haule, pp. 167–95, Dallas: Spring Publications, 1977.

Kaiser, P. K., 'Physiological response to color: a critical review', *Color Research and Application*, **9** (1984), 29–36.

Katz, D., *The World of Colour*, 1935 (reprinted London, 1970).

Kemp, M., *The Science of Art*, 2nd edn, 1991.

Kouwer, B., *Colours and their Character: a Psychological Study*, Gravenhage, 1949.

Kuschel, R., and Monberg, T., ' "We don't talk much about colour here": a study of colour semantics on Bellona Island', *Man*, **9** (1974), 213–42.

MacLaury, R. E., *et al.*, 'From brightness to hue: an explanatory model of color-category evolution', *Current Anthropology*, **33** (1992), 137–86.

Maerz, A., and Paul, M. R., *A Dictionary of Color*, 3rd edn, New York: McGraw Hill, 1953.

Nicholson, W., *Unknown Colour*, London: Faber, 1987.

Oxbury, J. M., Oxbury, S. M., and Humphrey, N. K., 'Varieties of colour anomia', *Brain*, **92** (1969), 847–60.

Sahlins, M., 'Colors and cultures', *Semiotica*, **16** (1976), 1–22.

Saunders, B., and van Brakel, J., 'On translating the World Color Survey.' In *Contemporary Anthropology*, ed. D. Raven and J. de Wolf, Assen, 1994 (in press).

Scott, I. (ed. and trans.), *The Lüscher Colour Test*, 1971.

Wierzbicka, A., 'The meaning of color terms: semantics, cultures and cognition', *Cognitive Linguistics*, **1** (1990), 99–150.

8   Colour in Language

John Lyons

Introduction: the vocabulary of colour

The title of my chapter can be interpreted broadly or narrowly, and, if narrowly, can have one focus rather than another. I am going to adopt a relatively narrow interpretation, and the focus that I will give it will be determined partly by the fact that I am communicating as a linguist to a readership that is composed for the most part of non-linguists and partly by the fact that this chapter appears in an interdisciplinary book that deals with, amongst other things, colour in Nature, colour in Culture (including colour in art), colour and light, and the neurophysiology of colour-perception. As we shall see, the topic, or set of topics, that I address under the rubric of 'colour in language' cannot be tackled sensibly without linking it up with topics dealt with in several of the other chapters, and the fact that my chapter comes last in the book greatly simplifies my task. Occasionally, I will make explicit the points of contact between what I am saying and what has been said by other contributors. But even when I do not make them explicit, it will generally be evident what these points are and in what respect they are relevant.

What I am going to discuss can be referred to non-technically as the vocabulary of colour in natural languages and, rather more technically in the jargon of my trade, as the *lexicalisation*, or lexical encoding, of colour in natural languages. I will avoid, as far as possible, the more specialised technical terminology that linguistics has developed in order to describe language and languages with greater precision than is possible by using, without qualification or re-definition, more or less ordinary English. I must make it clear at the outset, however, that it is very difficult to say anything sensible or significant either about language in general or about particular languages without re-formulating what is said in the technical metalanguage of modern theoretical linguistics or qualifying it parenthetically to the point of incomprehensibility with conditional and concessive clauses. For example, I shall be using the term *word*; and, throughout this chapter I write as if all natural languages have words. This is in fact incorrect unless the sense of the term 'word', which has several different meanings in everyday English, is tacitly adjusted or qualified either contextually or in relation to languages of different types. Those who have some background in linguistics will, I hope, make these adjustments and qualifications, as the occasion arises, and should have no difficulty in doing so, not only with respect to the term *word*, but also with respect to *language*, *sentence*, and so on. Those who do not are asked to accept my assurance that I could, if necessary, spell it all out – and indeed have done so elsewhere – at great length and with what many of my colleagues would say is extreme pedantry. All specialised disciplines, of course, suffer from the difficulty to which I have just referred: that of explaining satisfactorily in non-technical language what their practitioners are up to. But linguistics suffers from an additional disadvantage in this respect. Linguists, like everyone else, have to use language to talk about the world (and everything it contains); however, they also have to use language to talk about language and, what is of special concern to us, about the relation between language and the world. It is of importance therefore that they should be able to make it clear when they are referring to words, when they are referring to the meaning of words and, in so far as this differs from what they mean (in one sense of 'meaning'), when they are referring to things and properties in the world. For example, when I use the word **green** I might be referring to the English word **green** (or to its form independently of its meaning or vice versa) or to the colour it denotes. There is a whole host of more or less standard technical terminology and notation for making this clear, but I will rely primarily on context and on the tolerant and discerning comprehension of my readers.

Having said that I would avoid the use of jargon, I hasten to explain that my deliberate introduction of the terms *lexical*, *encode* and *lexicalisation* was not gratuitous. In present-day linguistics, the term 'lexicalisation' immediately evokes the complementary term *grammaticalisation*. The term lexical is related to *lexicon*, which for our purposes we may treat as being synonymous with 'vocabulary' or 'dictionary'. To lexicalise something, to encode it lexically in the language, is – to put it loosely but, I hope, comprehensibly – to provide a word for it in the vocabulary of that language. When I say, for example, that such-and-such a language has no word for brown, this is tantamount to saying that the colour brown or, alternatively, the concept 'brown', is not lexicalised in the language in question. Similarly, to say that a language has or does not have this or that grammatical category or grammatical distinction is to say that the language in question grammaticalises or does not grammaticalise that category or distinction. For example, there are many natural languages that do not grammaticalise the distinction of singular and plural, or of past, present and future, or of nouns, verbs and adjectives. It is important to realise that the encoding of distinctions of colour in particular natural languages is as much a matter of grammar as it is of vocabulary. This is something that is not as widely appreciated, even by linguists, as it ought to be. Most general discussions of colour in language do not even mention it. Regrettably, it is something that I cannot deal with here.

For simplicity of exposition, I shall write as if all languages had the same kind of grammatical structure as English or – to use Benjamin Whorf's evocative term, which is particularly appropriate in the present context – Standard Average European. It must be borne in mind, however, that English is but one of many thousands of natural languages spoken in the world today and, in various respects, including the encoding of colour, is highly untypical both grammatically and lexically of the languages of the world. But whatever individual language or small set of languages we use for the exemplification of general statements about language will be in various respects unrepresentative of the several thousand natural languages known to linguists, most of which have so far been very inadequately described.

What is colour? A linguist's view

Colour terms have figured prominently in twentieth-century linguistic theory and in the philosophy of language, especially in discussions of the relation between language and the world by linguists and philosophers. This is the general context within which I operate, although I cannot go into the philosophical issues in any detail. But I should perhaps state what view I am adopting on the major issues of contention in so far as they are relevant to my topic. My position is, I think, philosophically defensible, if not unchallengeable; much of what I have to say at particular points supports it with empirical evidence. Briefly, the philosophical approach that I am adopting is one of naive realism and moderate relativism. Let me explain what I mean, without, as I say, going into the details.

To say that I am adopting the doctrine of naive realism – duly qualified in a way to be explained later – is to imply that I am assuming that the objects of sense perception exist in the external world independently of the mind and of language, that they have real properties that are similarly independent of mind and language, and that the general terms such as *cow* or *table* applied to natural entities or artifacts of the same kind, such as cows or tables, respectively, are applied to them because they share the same essential properties – the properties that constitute their essence.

Now, colour as such is clearly not a real property in this sense. As other writers in the book have put it, colour is in the eye of the beholder; Denis Baylor (Chapter 4) and John Mollon (Chapter 5) between them, have explained brilliantly how the sensation of colour is induced by the processing by the eye and the brain of the electromagnetic radiation reflected off the entities and substances that I am assuming to exist. At the same time, nothing that they have said prohibits the naive realist from assuming that there are real external physical correlates to what are perceived as distinctions of colour; it is reasonable, if only to avoid circumlocution, to talk as if it is colour, rather than its physical correlates, that is the shared property of entities and substances which are said to be identical in colour. I will proceed, then, as if colour itself is a property of objects and substances that exist in the external world. Everything that I have to say about colour could be re-formulated, I believe, within the framework of a somewhat different, less naively realistic, set of ontological assumptions.

I am assuming, then, that colour is real. I am not assuming, however, that colours are real. On the contrary, the main burden of my argument is that they are not: my thesis is that they are the product of the lexical

and grammatical structure of particular languages. But let me not anticipate. For the present, it suffices for me to produce my naively realistic
definition of colour. This runs as follows:

> Colour is the property of physical entities and substances that is
> describable in terms of hue, luminosity (or brightness) and saturation and
> that makes it possible for human beings to differentiate between
> otherwise perceptually identical entities and substances, and more
> especially between entities and substances that are perceptually identical
> in respect of size, shape and texture.

This definition may seem, at first sight, to be unnecessarily cumbersome. But I have deliberately included within it a number of points that
will be of importance as we proceed. Let me draw your attention to them.
First of all, the definition I have adopted covers achromatic as well as
chromatic colours and gives no primacy to hue in contrast with luminosity and/or saturation: there are of course contexts in which the English
word 'colour' (and its equivalents in some other languages) is used more
narrowly.

Secondly, I have deliberately introduced into my definition of colour
the properties of shape, size and texture, as being, together with colour,
constitutive of the *appearance* of things: what they look like. Although I
shall not be dealing with the grammar of colour, I should mention that
all of these other properties are more widely grammaticalised in the languages of the world than colour is.

Thirdly, I have made explicit the fact (which emerges clearly from earlier chapters) that the perception of colour, and perception in general, is
species specific; that is to say, the way the world appears to members of
one species, more particularly to human beings, is specific to that species.
Language is similarly species specific. It has a biological, genetically transmitted, basis and, after the period of language acquisition, is stored as
one language rather than another, in the left hemisphere of the brain (in
most human beings). The species specificity of perception, cognition and
language, together with the degree to which particular natural languages
are the product of both Nature and Culture, is something with which we
have to reckon in any discussion of the relation between language and the
world. One of the organizing dichotomies in the chapters of this book has
been the dichotomy between Nature and Culture. I have less to say about

this from the viewpoint of linguistics than I would like, but I will say something about it as we proceed.

Finally, I should perhaps make explicit the fact that any realist definition of colour, as also of shape, size and texture, must be interpreted in the light of the well-established principle of perceptual constancy: in visual perception the brain automatically adjusts for and discounts (in normal circumstances) differences of perspective, distance and lighting when it judges two objects to be identical in size, shape and colour. I make this point explicit because there is a comparable principle, whose basis, however, may be to some considerable extent cultural rather than natural, with which we commonly operate when we talk, generically, about the properties of things: the principle of what I will call *prototypicality*. The reason for choosing this term will become clear in due course. For example, when we say that grass is green or that lemons are yellow, we do not mean that this is their colour in all instances and in all seasons: we mean that this is their colour prototypically or paradigmatically. The terms I have just used may be unfamiliar. The principle itself, I am sure, is not.

Let me now turn to the distinction between universalism and relativism. What is at issue in the present context is whether all natural languages have the same grammatical and lexical structure: whether all languages are grammatically and lexically isomorphic. Universalists say that they are; relativists say that they are not. Superficially, of course, it is obvious that few, if any, natural languages are grammatically isomorphic. For example, the fact that French, Russian and German have grammatical gender, whereas English does not, means that English is not grammatically isomorphic with any one of these three languages; the fact that gender operates somewhat differently in each of these three languages means that no one of them is isomorphic with either of the others. And so it is for literally hundreds of other features of the grammatical structure of these and other languages. Yet there are many philosophers and grammarians who have proclaimed themselves to be universalists rather than relativists. They have been able to do so, whether plausibly or not, because they have been concerned not with what they have regarded as the superficial structural properties of language but with what they have considered to be essential to it.

I do not wish to go into past or present theories of universal grammar, since these are not directly relevant, in their detail, to the vocabulary of colour. It suffices for the moment that a case can be made for the view that, although languages differ considerably in what I refer to as their grammatical superstructure, they have a common infrastructure. I am using these terms *superstructure* and *infrastructure*, loosely and quasi-metaphorically. But those who are familiar with contemporary linguistic theory Chomskyan and non-Chomskyan will be able to supply their own more technical interpretation.

When it comes to vocabulary, and more especially the vocabulary of colour, it is once again obvious that, superficially at least, not all languages are isomorphic. Let us consider just a few well-known examples to demonstrate the fact that the colour terms of particular languages cannot always be brought into one-to-one correspondence with one another: there was no word meaning 'brown' or 'grey' in Latin; literary Welsh has no words with the same meanings as the English words **green**, **blue**, **grey** or **brown**, but one word (**glas**) that covers part of green, another that covers the remainder of green, the whole of blue and part of grey, and a third term that covers the remainder of grey and part, or the whole, of brown; Japanese has a word, **awo**, which is translatable as 'green', 'blue' or 'dark' according to what it is used to describe; Russian has no single word meaning 'blue', the words **goluboi** and **sinii** (usually translated as 'light blue' and 'dark blue', respectively) denote what are in Russian distinct colours, not different shades of the same colour, as an English translation might suggest (and readers who have a good knowledge of Spanish or Italian will know that a comparable, though not identical, situation exists in those languages when it comes to words for what we identify as the colour blue); Hungarian has two terms for what we call red; Navaho has one word for both brown and grey and also a single word for blue and green, but two words for black; many languages of New Guinea and most Australian aboriginal languages have only two colour-terms meaning 'dark' and 'light' or, in appropriate contexts, 'black' and 'white'; and, as we shall see later, Hanunoo (a Malayo-Polynesian language of the Philippines) has just four basic colour-terms, which can be translated (in some contexts at least) as 'black', 'white', 'red' and 'green'. Examples such as these could be multiplied indefinitely and, a generation or so ago in the heyday of structuralism, were frequently cited in textbooks and monographs to support the validity of the principle of linguistic relativity: the principle that the structure of languages is wholly arbitrary and can vary unpredictably.

As we shall see presently, this conclusion has been challenged more recently, in a particularly interesting way, with respect to the vocabulary of colour, by Brent Berlin and Paul Kay. In my view, the work of Berlin and Kay has been widely misinterpreted and is far from having established as fact much that is now presented as fact in textbooks in linguistics, psychology and anthropology. But I will come to that in due course. What we have established so far is that a simplistic, or superficial, interpretation of the thesis that all languages have the same grammatical and lexical structure is untenable. Present-day universalists work with a less simplistic, less superficial, interpretation of their thesis. It is my major task in this chapter, first of all to expound this as far as the vocabulary of colour is concerned, and then to demonstrate that it is, though perhaps not wholly unfounded, less soundly based than current textbooks in linguistics and the psychology and philosophy of language might lead one to suppose. At the very least, the hypothesis of Berlin and Kay and the empirical results of their research and that of their followers need to be formulated with greater care than is customary.

English

I will now discuss a little more systematically the vocabulary of colour in particular languages. To get us started, I will make a few comments about colour-terms in English. I can then generalise these comments appropriately to other languages, qualifying my generalisations where necessary.

Just how many colour-terms there are in English it is impossible to say. Apart from the fact that it is not always clear what counts as a colour-term and what does not, we can define English more or less comprehensively, according to whether we decide to include or to exclude dialect variants and the specialised sub-vocabularies of artists, interior decorators, textile designers, paint manufacturers, and so on. A fairly comprehensive list would run into the hundreds, many of which would be unfamiliar to most native speakers of English.

Among the smaller number of words that definitely count as colour-terms, are in general use, and can be assumed to be known and understood by most, if not all, adult native speakers, we can distinguish up to a dozen or so common level-1 terms, including **black**, **white**, **red**, **green**, **yellow**, **blue**, **brown**, **purple**, **pink**, **orange** and **grey**, and a larger number of level-2 terms, such as **scarlet**, **mauve**, **turquoise** and **beige**. This distinction between level-1 words and level-2 words is, I trust, both intuitively obvious and uncontroversial. Level-1 words are more general; level-2 words are more specific. Level-1 words are those that would normally be used in answer to the question 'What colour is X?', if there is no need in context to be more specific: they are for that reason much more frequent in everyday discourse and are learned earlier by children in the acquisition of their native language. They are therefore, in this sense, more basic than level-2 words. Level-2 words are also less basic in another sense: they can often be defined, more or less satisfactorily as the case may be, in terms of level-1 words. For example **mauve** might be defined as '[a particular kind of] pale purple', **scarlet** as 'a brilliant red (tinged with orange)', **beige** as 'a yellowish grey' or, alternatively, 'a pale dullish yellowish brown', and so on (these definitions, or glosses, come from recent authoritative dictionaries of English).

The distinction between level-1 and level-2 words rests upon one of the principal structural relations in the vocabularies of natural languages: the relation of inclusion (or, more technically, hyponymy). This can be plausibly accounted for, historically, not only with respect to the vocabulary of colour but, more generally, in terms of the interplay, over time, of a combination of so-called pragmatic principles, having to do with communicative efficiency. The most important such principle in the present case is what linguists and philosophers will recognise as a version of Paul Grice's maxim of quantity: 'Do not be more specific than you have to be in the context in question and for the purpose in hand'.

A second structural relation that holds among the level-1 words (and within some subsets of higher-level terms) is that of incompatibility. For example, **red** and **blue** are incompatible in the sense that something cannot be wholly red and wholly blue. True, one or two tough-minded philosophers have challenged the validity of this semantic postulate or axiom. But most philosophers of language have taken it to be valid (though they may have argued as to whether it is analytically true or not); and so shall we.

A third structural relation that holds between pairs of level-1 words is that of adjacency and, as a particular case of adjacency, intermediacy (or between-ness). For example, **orange** is adjacent to both **red** and **yellow** and, more particularly, is between them both. Obviously, if there is more than a single dimension of variation – hue, saturation and luminosity – x can be adjacent to both y and z without being between them. As it happens, all three psychophysical dimensions of variation need to be invoked in specifying the denotation of several of the level-1 colour terms of English. This is not the case in all languages. The structural relation of adjacency is reflected in the possibility of using the productive resources of the language to form compound or complex level-1 terms such as *blue-green*, on the one hand, or *reddish-brown*, on the other, without violating the axiom of incompatibility.

Two further points should be made before we proceed. The first is that English has several words that may be used to qualify or modulate level-1 colour-terms (although this is not their sole or primary use). They include such adjectives as **deep, pale, light, dark, bright, vivid** and **brilliant**. The definition of level-2 words in terms of level-1 words – e.g. the definition of **mauve** as 'pale purple' or **scarlet** as 'brilliant red' – makes essential use of the productive grammatical and lexical resources of English that make it possible to identify and define, without creating new higher-level colour-terms, indefinitely many additional words and phrases of increasing specificity.

The second point to be made is that there is no level-0 adjective in English superordinate to the level-1 adjectives (and for the moment let us treat them as adjectives) that I have identified; that is to say, there is no adjective in general use in English that means 'having colour' and serves as a true antonym, or opposite, to *colourless* in its broadest sense. Both *coloured* and *colourful*, as they are generally used, have a somewhat different meaning from 'having colour'. (Admittedly, *coloured* is sometimes used in philosophical discussions to mean 'having colour'; but this is a recognisably specialised usage.) English is by no means unique in this respect. Nor is English unique, or indeed in any way untypical of the languages of the world, in respect of the structural relations that I have mentioned or of the use that is made of them in the organisation of the vocabulary of colour in those languages that have – and, as we shall see, many do not – a reasonably rich colour-vocabulary.

At this point, I should emphasise that all the structural relations that I have mentioned are language-internal and can, in principle, be learned without having learned in advance or without learning simultaneously the denotation of the terms in question that links the language to the world. Not only is this so in principle; it is also demonstrably so in practice. Blind people can successfully learn the language-internal meaning of colour-terms. This fact, incidentally, invalidates the strongly empiricist theory of semantics according to which the meaning of such allegedly basic words as 'red' or 'green' is accounted for wholly in terms of the link between the word and the property or class of objects that it denotes and can be learned by so-called ostensive definition without one having learned in advance or simultaneously at least some of their language-internal meaning.

The list of level-1 colour-terms that I gave earlier contained eleven words. I might have added one or two more, but, as we shall see presently, there is a particular reason why I have restricted myself to just these eleven: these are the eleven words that are regarded as 'basic' in much recent work influenced by Berlin and Kay. We can group them into several subclasses. Three of them are achromatic: **black**, **white** and **grey**. Of the eight chromatic colours, five are what I will call Newtonian: i.e. they are included in the now traditional list of seven prismatic, or spectral, colours established by Newton in his *Optics* and can be thought of as having hue as their principal dimension of variation. These are **red**, **orange**, **yellow**, **green** and **blue**. (Newtonian **indigo** and **violet** are missing, their place being taken by **purple**.) **Brown** and **pink** are clearly *non-prismatic*, in that luminosity and saturation (or rather desaturation), respectively, are essentially involved in their definition.

When it comes to the definition of the level-1 colour vocabulary in standard dictionaries of English, it is not uncommon for some words to be defined in terms of other level-1 words in much the same way that level-2 words are defined in terms of level-1 words. For example, *green* might be defined as being between **yellow** and **blue** (or, alternatively, as denoting a colour that results from the mixture of yellow and blue); **purple** might be defined similarly in terms of **red** and **blue**; and so on. It is noticeable, however, that there is no general agreement as to which of the level-1 words are in this sense definitionally more basic, or more primitive, than the others. It is not the case, for example, that the words for the colour-theorists' primary colours (**red**, **yellow** and **blue**) are invariably chosen as being definitionally more basic than words that denote their complementaries. And it is rare for lexicographers to invoke the notion of antonymy, or oppositeness, except in respect of **black** and **white**. It is also worth noting in this connection that semanticists who have espoused an atomistic, or componential, theory of lexical meaning based on the notion of binary opposition have rarely applied this theory to the vocabulary of colour, despite the fact that it has some *prima facie* validity in the neurophysiology of colour perception. Facts of this kind, and others that might be adduced, should make us wonder whether all those who operate professionally with colours and colour-terms – artists, physicists, physiologists, linguists, and so on – can be expected to agree that certain colours or colour-terms are in all respects and for all purposes more basic than others. Let us keep this in mind when we consider, as we shall be doing shortly, the influential theory of basic colour-terms developed by Berlin and Kay. But let us now take a quick look at French.

French

French, like English, has a rich multi-level vocabulary, and all the generalisations I have made so far about English hold also for French. One of the most authoritative and comprehensive dictionaries of French (*Le Grand Robert*, 1966) lists under the word *couleur*, conveniently for our purposes, almost 200 'principales couleurs', each of which is defined in terms of one of eleven French level-1 words. All but two of these can be put into correspondence with one of the eleven level-1 words of English. The two exceptions are **violet**, which arguably does not quite match English **purple** (French **pourpre** is the equivalent of the English **purple** only when the latter means 'royal purple'), and **brun**, which certainly does not match the English word **brown** in most contexts in present-day French.

The status and meaning of the word **brun** in the vocabulary of modern French can be used to illustrate a principle of considerable importance, which will enable us to make sense of what is at first sight paradoxical, or problematic, in the colour vocabulary of many other languages. There is no doubt that **brun** is a level-1 word in present-day French, both literary and colloquial. It may also still be regarded as basic (though there is some evidence that, in this respect, it is yielding ground to **marron**) in the sense that, when one is referring to colours as such, it can be used with such qualifying words as **clair** ('light') or **foncé** ('dark') to define other colours. For example, **beige** (which arguably, however, is a level-1 word in French, as the English **beige** clearly is not) can be defined as **brun très clair**. Moreover, when **brun** is used to refer to a particular colour, in abstraction from things that have the colour in question, it seems to be more or less denotationally equivalent to – in this sense therefore, to have the same meaning as – the English **brown**. But, as anyone who knows French will be aware, the situation is very different when it comes to the *descriptive* or *attributive*, rather than the *referential*, use of **brun**: that is to say, when one is not referring to the colour brown itself, but to objects or substances that are brown in colour and either identifying or describing them in terms of their colour.

This distinction between reference and description is crucial in semantics. Failure to draw it properly has led to a good deal of confusion in linguistics and in the philosophy of language. In the present instance, the difference between the two becomes clear as soon as one reflects upon the two interpretations of the ambiguous English utterance: 'That's brown', paraphrasable as either 'That colour is brown' or 'That object (or substance) is brown'. The fact that this utterance is ambiguous in this way depends, in particular, on the grammatical ambivalence of colour-terms in English.

As for **brun**, first of all, in many contexts one would not use it at all. For example, one would not use **brun** of brown shoes, but **marron** (or perhaps **beige**, or even **jaune**, for a lighter brown). Secondly, when it is used of the colour of someone's hair, skin or eyes, it means 'dark', rather than specifically 'brown'. (The relation between **brun** and **marron** has been investigated in several recent works, and supports what I am saying here.) In one or other of these two respects, both of which can be subsumed under the term *context-restricted*, **brun** is typical of many level-1 colour-terms in certain languages, and I shall make this point with particular reference to Ancient Greek presently.

Context-restrictedness is perhaps even more characteristic of higher-level, or stylistically more specialised, colour-terms, in any language which has a rich colour vocabulary. For example, among the 180 or so higher-level or more specialised 'principales couleurs' listed in *Le Grand Robert*, the dictionary to which I referred above, we find the delightfully evocative **cuisse de nymphe** (a shade of pink, literally translated as '[colour of a] nymph's thigh') and the perhaps less delightfully evocative **caca d'oie** (defined as a kind of **brun**, literally translated, in a somewhat more vulgar register of English, however, as 'goose shit'). These two terms are, of course, part of the more or less specialised vocabulary of painters and art historians, and are indeed not infrequently used in English. But what I want to draw attention to is the fact that several of them are not so much stylistically restricted (though they may be obsolete or obsolescent and unfamiliar to many native speakers of French) as restricted by what they can be used to describe. For example, no less than ten are said to be used mainly or exclusively to describe horses (rather as **bay** and **sorrel** are used in English), one is said to be used of human hair (**auburn**), one of leather, and so on. No less interesting, however, than the inclusion of a word such as **auburn** is the exclusion of words such as **blond** (used characteristically of hair, beer and tobacco).

There are two points to be made on the basis of examples like this, which can of course be matched in English, German, Russian, or, I imagine, any language that has a rich colour vocabulary. The first is that, in any linguistically significant sense of 'basic', the fact that a word is context-restricted should not be thought to make it less basic than one that is not (though it may, of course, make it less basic in some philosophical sense of the term). The second is that one has to be very careful to draw a distinction between the referential and the descriptive use of colour-terms. The term *colour-term* itself turns out to be ambiguous; its ambiguity gives rise to the paradox that it may be both true and false to say of particular words in particular languages that they are or are not colour-terms. This second point is particularly important, and I shall exploit it subsequently. To the best of my knowledge it has not been given the attention it deserves in recent discussion of colour vocabulary by linguists and others.

The Berlin–Kay hypothesis: do all languages recognise the same (linguistically) basic colours?

Much of the research done on the vocabulary of colour in natural languages over the last twenty years or so has been stimulated by Berlin and Kay's interesting and influential attack on the relativistic view of colour-terms associated with structural linguistics: *Basic Color Terms: Their Universality and Evolution.* I have already referred to this. It has also been mentioned by John Gage, somewhat critically, in Chapter 7.

I will be equally critical, not of the Berlin and Kay hypothesis itself, which I will henceforth refer to as the BK-hypothesis, but of popularisations of it, which, to my mind, are based on misunderstanding and misinterpretation and on a failure to appreciate that there is a crucial ambiguity in the term colour-term that promotes misinterpretation of their results and those of their followers. I could also criticise quite severely on methodological grounds, as others have done, some of the earlier work. But I prefer to concentrate on what, I think, is both more original and more relevant to our present topic. I will begin by providing a brief summary of the BK-hypothesis for the benefit of those who are not familiar with it.

According to Berlin and Kay, all natural languages have between two and eleven basic colour-terms. These are (to use English as our metalanguage): **black, white, red, yellow, green, blue, brown, purple, pink, orange** and **grey**. (This list, it will be noted, is identical with the list of level-1 English colour-terms that I listed earlier, as now, in their BK-hierarchical order. The significance of the term *BK-hierarchical* will be evident presently. English is one of the languages that is said to have the maximum of eleven basic colour-terms.) What Berlin and Kay mean by 'basic' – 'BK-basic' – depends upon the application of several criteria. The four most important are as follows: (1) a basic colour-term must be lexically simple (i.e. it must not be morphologically or syntactically composite, as are **reddish, red-brown, honey-coloured, like grass** [in colour], etc.); (2) it must be a level-1 term; (3) it must be psychologically salient and in common use; (4) it must not be context-restricted. All but one of these, in my view, are both linguistically justifiable and relevant to the purpose. The fourth, as I have suggested earlier with respect to the French **brun**, is not.

But for the moment let us take on trust the composite notion of 'BK-basic' and assume that it is both valid and reliable. On the whole, I think that this is so (provided that it is understood that it relates to the referential, rather than the descriptive, use of colour vocabulary). It correlates well with such measures as frequency of use, saliency of recall, immediacy of response, and in particular cases yields results that are consistent with the intuitions and judgements of experienced lexicographers.

What Berlin and Kay maintain is: that there are eleven psychophysically definable focal points, or areas, within the continuum of colour and that there is a natural hierarchy among at least six of these focal areas determining their lexicalisation in any language; that all languages with only two basic colour-terms have words whose focal point is in the area of black and white (rather than, say, in yellow and purple); that all languages with only three basic colour-terms have words for black, white and red; that all languages with only four basic colour-terms have words for black, white, red and either green or yellow; that all languages with only five basic colour-terms have words for black, white, red, green and yellow; that all languages with only six basic colour-terms have words for black, white, red, green, yellow and blue; and that all languages with only seven basic colour-terms have words for black, white, red, green, yellow, blue and brown. As to the remaining four focal areas, purple, pink, orange, grey, there is no hierarchy among these: no one of them takes precedence over the others. This then is the BK-hypothesis. It has been pointed out by E. V. Clarke and H. H. Clarke that 'if combinations of the eleven basic colour terms were random, there would be 2048 possibilities' and that the BK-hypothesis 'restricts that number to thirty-three'.

Associated with the BK-hypothesis, there are two so-called evolutionary sub-hypotheses: the one *phylogenetic*, the other *ontogenetic* (as these terms, borrowed of course from zoology, are nowadays used in linguistics). The phylogenetic hypothesis is that, in the historical development of languages over time, whenever languages acquire new basic colour-terms, they will always acquire the next one down the hierarchy. This means that languages can be grouped, from an evolutionary point of view, as: stage-1 languages, with only two colour-terms, denoting what I will call BK-black and BK-white; stage-2 languages with only three basic colour-terms, BK-black, BK-white and BK-red; stage-3 languages with only four basic colour-terms, BK-black, BK-white, BK-red and either BK-green or BK-yellow; and so on. It also means that there can be no natural language that has, let us say, basic colour-terms for only the BK-colours black, white and green or for only BK-red, BK-green and BK-blue. As to the ontogenetic sub-hypothesis, which has to do with the learning of colour-terms by children in the acquisition of their native language (this was put forward much more tentatively by Berlin and Kay), this is that they first master the distinction of the BK-colours black and white, then learn BK-red, afterwards BK-green or BK-yellow, and so on.

Various minor revisions of the original BK-hypothesis have been made over the years, and (as is explained in the references listed under further reading) several facts about particular languages have been cited that would tend to invalidate a very rigid interpretation of the hypothesis, whose effect, as we have seen, is to set the absolute upper limit for possible combinations of basic colour-terms at thirty-three. For example, in many languages a term for BK-grey developed earlier than was originally proposed; BK-brown is a problem in Russian, and, as was noted earlier, also in French; and Russian has two basic words for blue, one of which appears to have developed before the term for purple. However, the general opinion seems to be that the BK-hypothesis is essentially correct. For present purposes, therefore, I will assume that it has been to some degree confirmed by further research over the last twenty years and has certainly not yet been falsified. What conclusions follow?

The most important things to note in the present context is that a distinction has to be drawn between the central, or focal, denotation of a word and its total denotation. Two languages might well differ with respect to the boundaries that they draw in a denotational continuum and yet be in agreement with respect to what is central, or focal, in the denotation of roughly equivalent words. For example, granted that the literary Welsh word **glas** overlaps the areas that are referred to in English as green and brown, it may well turn out to denote what I am calling BK-blue: i.e. a colour whose focus is that identified for blue by Berlin and Kay; and, interestingly enough in this particular case, this appears to be its principal meaning in modern spoken Welsh, though it is still used for the colour of vegetation. Independently of the ultimate fate of the BK-hypothesis or some replacement of it, there is no doubt that the idea that languages might be in agreement about the focal area of a colour – about what is, let us say, a prototypical blue or green – and yet differ about the number of basic colour-terms in the system or about where the boundary comes between two adjacent colours has been enormously influential and productive of much interesting research. So much, then, for my summary of the BK-hypothesis. Let us now look at the basic colour-vocabulary of a less familiar language than English and French in the light of the BK-hypothesis.

Hanunoo: an exotic and strikingly different language?

What I am now going to present in some little detail are the results of research carried out by the American anthropologist Harold Conklin in the early 1950s, more than fifteen years before the seminal work of Berlin and Kay was published and a decade or so before Chomsky's theory of universal grammar began to have the effect that it eventually did have in changing the prevalent view of linguists from relativism to universalism.

Conklin's analysis of the vocabulary of colour in Hanunoo – the Malayo-Polynesian language of the Philippines mentioned above as having four colour-terms – is one of the classics of ethnolinguistic or anthropological structural semantics. Regrettably, it is not as well known nowadays as it ought to be, being research of a kind that many linguists, psychologists and anthropologists wrongly believe to have been superseded by that of Berlin and Kay and their followers. Conklin's findings, as we shall see, are wholly consistent with the BK-hypothesis. At the same time they point to the fact that the BK-hypothesis is far less relevant to the analysis of colour-terms in some languages than it is in others. Rather more surprisingly, Conklin's findings can be interpreted as throwing light on the analysis of colour-terms – or so-called colour-terms – in such familiar and well-studied languages as Ancient Greek and, to a lesser extent, Latin.

Hanunoo, as we have noted, has four level-1 colour-terms (and several hundred higher-level terms, which can be used 'when greater specification than is possible at level-1 is required'), which denote areas of the colour continuum whose foci are in black, white, red and green. As

Conklin himself puts it: 'in general terms, **mabi:ru**, includes the range usually covered in English by black, violet, indigo, blue, dark green, dark grey, and deep shades of other colours and mixtures; **malagti?** white and very light tints of other colors and mixtures; **marara?**, maroon, red, orange, yellow and mixtures in which these qualities are seen to predominate; **malatuy**, light green and mixtures of green, yellow and light brown'. These four terms can be satisfactorily identified with the BK-colours black, white, red and green (i.e. BK-black, BK-white, BK-red and BK-green) despite the relatively large area of the colour continuum that each of them covers, because 'the focal points [within each area] can be limited more or less to black, white, orange-red, and leaf-green respectively'.

So far, so good. Hanunoo is clearly a well-behaved BK-stage-3 language, of which there are several other examples referred to in the literature, and it was classified as such by Berlin and Kay in 1969. What is not pointed out, however, is that, according to Conklin, although the four words we have been considering can be called colour-terms and could no doubt be elicited as such – by clever and somewhat laborious questioning (there is no word for colour in Hanunoo) – in laboratory tests with standardised colour-chips, chromatic variation does not seem to be the basis of their differentiation. The two principal dimensions of variation are lightness versus darkness, on the one hand, and, on the other, wetness versus dryness, or freshness (succulence) versus desiccation. To quote Conklin: 'A shiny, wet, brown-coloured section of newly-cut bamboo is **malatuy** (not **marara?**)' whereas 'dried out or matured plant material – such as certain kinds of yellowed bamboo or hardened kernels of mature or parched corn are **marara?**' There is also a third dimension of variation that divides the four terms, two by two, as 'deep, unfading, indelible', on the one hand, and 'pale, weak, faded, bleached or "colorless" '. But I will say no more about this, except to emphasise that the way it interacts with and cross-cuts the other two dimensions strengthens the general point that Conklin makes, and which I accept, about the interdependence of the Hanunoo colour vocabulary and Hanunoo culture.

At first sight, the colour vocabulary of Hanunoo, as analysed by Conklin, might appear to us to be exotic in the extreme, especially in its failure to separate hue, or chromaticity, from texture and succulence. And we might even be inclined to attribute this feature of Hanunoo colour-terms to the fact that the language operates in a culture which is in some sense more primitive, or less advanced, than more familiar Western European cultures. But is this feature of Hanunoo colour-terms really so exotic or unusual? It is my contention that it is not, and to demonstrate that it is not I will now present a partial analysis of the colour vocabulary of Ancient Greek.

Ancient Greek: is it so very different from Hanunoo?

Scholars have long been aware of problems attached to the translation of particular colour-terms in the Classical languages. They have argued about whether Latin **purpureus** meant 'red' or 'purple' or something different; they have wondered what shade of blue the Latin word **caeruleus** denotes and whether it could ever mean 'blue eyed'; they have puzzled over the true meaning of the Greek word **glaukos** (which was used most famously, of course, of the eyes of Athene); more generally, they have noted the impossibility of translating the colour-terms of Greek and, to a lesser extent, Latin consistently and systematically into English, French, German or any modern European language. So acute is this problem that some scholars have seriously considered whether the ancient Greeks were colour-blind. Only a few years ago in a collection of plays in simple Latin intended for school children learning the language (appended to a quite reasonable statement to the effect that the adjective **purpureus** was used 'for a bright and impressive colour which we might call purple or red') I noted the comment: 'Romans were not very good at distinguishing colours'. This is, to say the least, a misleading generalisation; literally interpreted, it is unquestionably false. All the evidence currently available suggests that our ability to distinguish shades of colour in terms of variations of hue, luminosity and saturation (in standard laboratory conditions) is uninfluenced by the structure of the colour vocabulary in our native language. This means, incidentally, that what is sometimes referred to as the strong version of the Sapir–Whorf hypothesis of linguistic relativity and linguistic determinism has been falsified. But it is not clear that Edward Sapir or Benjamin Lee Whorf ever subscribed to this interpretation of the hypothesis that bears their name. More moderate versions of the Sapir–Whorf hypothesis, of the kind to which they

213

themselves might well have subscribed, have never been falsified and can indeed be said to have been confirmed by a good deal of empirical research, including research on the vocabulary of colour.

It is impossible here to discuss fully the structure of the vocabulary of colour in Greek and Latin. There are two reasons, however, why I should say something about these two languages. The first reason, as I have already indicated, is that both of them to some extent, and Greek more than Latin, bear a striking resemblance to Hanunoo (and many other languages) in precisely those features that strike us, at first sight, as exotic (although the fact that this is so is perhaps not as widely known, even to classicists, as it should be, and, to the best of my knowledge, has not been mentioned in general discussions of the vocabulary of colour by linguists). The second reason is that much of the philosophical and scientific discussion of colour until modern times was based on ideas that go back to classical times and had been first expressed and illustrated, necessarily, in Greek and Latin. Malcolm Longair (Chapter 3) explained, and this point was picked up and emphasised by John Gage (Chapter 7), that the reason why Newton identified seven prismatic colours, rather than some smaller or larger number, was that he recognised an analogy between colour as it is apprehended by the sense of sight and sound as it is perceived by the sense of hearing. This analogy goes back ultimately to the Greek philosophers, and, before embarking on the analysis of the Greek vocabulary as such, it may be worth while introducing a short digression intended to link the scientific chapters in this book with the cultural and also to illustrate in advance one of my major points, that the colour vocabulary of particular languages is in part the product of culture (of which science, ancient and modern, is itself also a product).

One of the two most influential theories of colour among the Greeks, originated with – or at least was made famous by – the fifth century B.C. philosophers Empedocles and Democritus, who associated the four primary colours with the four elements: earth, air, fire and water. It is not without interest in the present context that these four primary colours can be identified with the BK-colours black, white, red and green or yellow. This association of the primary colours with the elements was incorporated by Democritus in his more general atomic theory (which included a corpuscular theory of light and vision) and also, independently of its association with a particular version of atomism, became an essential part of alchemy. I mention this point here because of Newton's interest in alchemy, which he took no less seriously than his interest in physics. As we have noted, Newton's analysis of the spectrum into seven chromatic colours – the seven colours of the rainbow as we refer to them in everyday parlance – was influenced by his commitment to the traditional analogy between musical and optical harmony. In this respect, as in his alchemy, he was heir to a tradition initiated by the Pythagoreans, arguably the first mathematical physicists and until modern times the most thoroughgoing, which extended what we see as the analogical principle of harmony (but which they took literally) into every domain of what we now call science, including sensory perception in every modality. As far as colours were concerned, harmonious combinations of the primaries produced pleasant colours, dissonant combinations produced unpleasant colours. And, as many commentators have noticed, both the notion of producing indefinitely many colours (some harmonious and some dissonant) by mixing primaries and also the choice of the four primary colours (by the atomists) was supported, if not caused, by the practice of the Greek painters of the period. If space allowed I would have said rather more about this in the final section of my chapter, where I shall be emphasing the interdependence of language and culture. As it is, this digression may perhaps stand as a small part of what could be, and should be, a long appendix exemplifying this interdependence.

A good starting point for any discussion of Ancient Greek colours is W. E. Gladstone's paper, 'Homer's perceptions and use of colour', published in the third volume of his monumental *Studies on Homer and the Homeric Age.* (It is not widely known, I think, that Gladstone was a very distinguished classical scholar and published a large number of articles and monographs, as well as translations of both Greek and Latin authors.) The main thrust of Gladstone's paper is to demonstrate that the Homeric concept of colour was less mature and more 'indefinite' than ours. The evidence that he marshalled in support of this thesis included: the paucity of colour terms; their infrequency of use by Homer for poetic effect 'where we might confidently expect to find [them]'; 'the vast predominance' of what he called 'the most crude and elemental forms of colour ... and the tendency to treat other colours as simply intermediate modes between these extremes'; and, as he put it, the use of the same word 'to denote, not only different hues or tints of the same colour, but colours which, according to us, are essentially different'.

There is much in Gladstone's formulation of his thesis that one would call into question 150 years later: for example, his apparent assumption of the primacy and reality of Newton's seven prismatic colours (at least three of which he found to be unrepresented in both Homeric and Classical Greek); and, of course, his too ready assumption that such facts as the paucity of colour-terms, the indefiniteness of their denotational boundaries, their infrequency of use, etc., reflected what we would describe as a psychological, or neurophysiological, difference between 'the Greeks of the heroic age' and men and women of his own day whose 'organ of colour' was more highly developed – the British having 'some special aptitude in this respect, [as] we may judge from the comparatively advantageous position, which the British painters have always held as colourists among other contemporary schools'.

Some of Gladstone's assumptions are no longer as tenable as they were in his day. But his evidence is for the most part sound; and his general conclusions have stood the test of time. Of particular relevance to my theme is his claim that of the more than twenty words in Homer that are commonly translated as colour-terms the majority are not basically colour-terms at all and those that are – only eight in number according to Gladstone – are understood as describing things primarily in terms of luminosity, or brightness, rather than hue.

Writing almost seventy years later, Maurice Platnauer, in a short but important paper, came to much the same conclusions for post-Homeric, Classical, Greek:

> Many colour epithets are not purely colour epithets at all, but have another meaning, and that meaning often enough is not even visional ... [What] seems to have caught the eye and arrested the attention of the Greeks is not so much the qualitative difference between colours as the quantitative difference between colours. Black and white are 'colours', and [other] colours are accounted for as shades between these extremes. It follows from this that no real distinction is made between chromatic and achromatic: for it is lustre or superficial effect that struck the Greeks and

not what we call colour or tint ... This may come from one of two causes: either the Greeks were definitely colour blind, or at least the colours made a much less vivid impression upon their senses ..., or, as I think is more likely, they felt little interest in the qualitative differences of decomposed and partially absorbed light.

Let me now note just a few further points relevant to the BK-hypothesis and to my earlier assertion that there are parallels between Hanunoo and Ancient Greek.

I do not think it is possible to establish a definitive list of BK-basic colour-terms for Ancient Greek, Homeric or Classical, in the way that Berlin and Kay and their followers have done for a large number of modern languages. We certainly cannot follow the example of Odysseus and, going down to Hades, tempt with a bowl of blood a representative sample of native speakers to label particular areas of the standard Munsell colour continuum with the most salient and the most appropriate everyday colour-terms! The best we can do is to put bits and pieces of evidence together and to interpret it as being indicative of how the Greeks themselves ranked or grouped the terms that they subsumed under the word *chroma* (or *chroia*), which we translate as 'colour' – and, no doubt, correctly in terms of the definition of colour with which I am operating. When we do this, especially if we rely mainly on passages from the Greek philosophers who have a commitment to a particular theory of colour, we must be careful not to take the evidence at face value without comparing it with other evidence that is available to us: it is at least possible that the terms that are grouped together or ranked more highly than others have been selected, deliberately or not, because they support a particular theory.

As we have noted, both Gladstone and Platnauer came to the conclusion (as did Conklin for Hanunoo) that luminosity is more important than hue in the colour vocabulary of Greek. In coming to this conclusion, neither Gladstone nor Platnauer drew essentially on the works of the philosophers. But there are many well-studied passages by Plato, Aristotle and others that strongly support the view that it was more natural to the Greeks than it is to us to think of the basic colour-terms (whatever they are and however many) as being arrangeable on a scale between **melas** and **leukos** at the end points – **melas** and **leukos** being translatable into English either as 'black' and 'white', or 'dark' and 'light', according to context. Most of the commentators on the passages in question agree that this is so. Several commentators also accept that, despite what standard dictionaries of Greek might say or imply, **melas** and **leukos** do not each have two meanings: the distinction between 'black' and 'dark', on the one hand, and between 'white' and 'bright', on the other, is an artifact of the process of translation into a language of different structure and is often a matter of arbitrary decision on the part of the translator.

Greek also has other words that are translatable as 'bright' (or 'light') and 'dark'; and when these are used in contrast with **leukos** and **melas**, especially in philosophical discussions of colour, they can be said to induce, in context, what we can then legitimately identify as the focal, or prototypical, denotation of **leukos** and **melas**: namely, BK-white and BK-black. The point that I have just made in relation to the Greek words **leukos** and **melas** applies, as we shall see, to other colour-terms, not only in Greek, but in other languages, and (though this is of no immediate concern to us) in other areas of the vocabulary. It expresses a principle that is of very general import in semantics; when we apply it, taking account of the distinction that I drew earlier between the referential and the attributive use of colour-terms, not just to the Greek words **leukos** and **melas** but to colour vocabulary throughout the languages of the world, we are well on the way to resolving the problems, or pseudo-problems, that linguists and translators encounter, if they operate with the traditional, simplistic, notion that each word in a language has a fixed number of one or more separate (but related) literal meanings, each of which has fixed and sharply drawn, rather than somewhat fuzzy, boundaries.

From what has been said so far, it is clear that the Greek **leukos** and **melas** are comparable with two of the four Hanunoo level-1 colour-terms; and, like them, pose no problem for the Berlin and Kay hypothesis. But what of the next highest colours in the Berlin and Kay hierarchy? There is no doubt that Ancient Greek, both Homeric and Classical, lexicalised the BK-colours red, green and yellow, in the sense that it had words that could be used to refer to all three, each in contrast with the others. Moreover, subject to the reservation that I expressed earlier about not taking at face value the evidence of philosophers theorising about colour, we can be fairly confident that, if we were in a position to carry out an investigation for Ancient Greek of the kind that has been carried out for a large number of modern languages by Berlin and Kay and their followers, the BK-colours red, green and/or yellow would emerge as more basic than any of the other colours lexicalised in Ancient Greek apart from BK-black and BK-white. And this would, of course, be as predicted by Berlin and Kay for any language higher than stage 4 in their evolutionary hierarchy.

Let us grant, then, that Ancient Greek had words for red, green and yellow. The problem is that it had more than one word for both red and green, no one of which is obviously a context-independent level-1 word: i.e. a more general word to which the others are, in all contexts, subordinate (or hyponymous), as **scarlet** and **crimson** are subordinate to **red** in English. Most classicists, if asked, would probably say that the basic, or general word for red is **eruthros** (which, as it happens, is etymologically related to what is certainly the basic word for red in many Indo-European languages, ancient and modern, including of course English, German and, via Latin, the Romance languages (French, Italian, Spanish, Portuguese, etc.)), and that such words as **phoinikous** and **porphurous** are indeed subordinate to it, exactly as **crimson** and **scarlet** are subordinate to **red** in English. This view is not wholly erroneous. There are passages in which **phoinikous** is used with a more specific meaning in explicit or implicit contrast with **eruthros**. But there are others (notably in Aristotle) where it is used to refer to one of the four most basic colours; and there are some passages in which it alternates with **porphurous**. Although there are passages in which **phoinikous** or **porphorous** have a more specific meaning than **eruthros**, there are others where they do not. And this is in accord with the principle that I introduced in relation to the words for black and white: induction of a narrower context-dependent meaning by contrast with what is in other contexts a synonym. Any one of these three words can be used to refer more generally to what we can reasonably assume to be BK-red.

At this point I should perhaps introduce another digression, to link what I am saying here with points made by other contributors to this book. It has to do with the etymology of the English word 'purple' and related words in other languages. The English word **purple**, like the

French **pourpe** and the German **Purpur** – all three of which, incidentally, differ from one another in meaning or level – derives, via Latin, from the Greek word **porphurous**. The history of this word and its derivates in various European languages illustrates beautifully the importance of cultural factors in the development of what has come to be in some languages, though not others, a basic level-1 term in the vocabulary of colour. As Peter Parks mentions (Chapter 6), what we call royal purple (which is not of course a particular shade of what we nowadays call purple in English, but like the Latin **purpureus** and the French **pourpre**, denotes a brilliant red) is the colour of the dye obtained from the shellfish *Pura haemastoma* or *Murex truncules*, and I should perhaps add that it took 12 000 of these shellfish to produce 1.5 grams of dye. The English words **purple** and **indigo** have a roughly similar history, in that the form of each word reveals the geographical origin of the dye to which the words from which they derive originally referred, and both words eventually became colour-terms. They differ, however, in that **purple**, unlike **indigo**, has for some centuries been a level-1 colour-term. Regrettably, I have not been able to go into the etymology of colour vocabulary, in English or other languages. Much of it derives ultimately from the prototypical colour of some culturally important metal or pigment; the higher-level colour vocabulary in European languages has been greatly extended in the post-Renaissance period by the adoption of terms originally employed by artists.

Much the same holds for the Ancient Greek words for BK-green as it does for the words for BK-red. There are several of these, no one of which is obviously more basic than the others. They include **chloros**, **prasinos** (or **prasios**) and **poodes** (and others that I do not discuss here); when they are being used referentially, and non-contrastively, as colour-terms, they do not seem to differ in meaning.

Of particular interest to us here is *chloros*, which is used more particularly of plants and foliage. It was this word that I had in mind when I said that Ancient Greek was similar to Hanunoo. Standard dictionaries of Greek will say that *chloros*, like the English word *green*, has two meanings, in one of which it denotes a colour and in the other of which it can be paraphrased by such words as *fresh*, *unripe* or even *moist* and *full of sap*, according to context. But this is not so. The colour-term sense of **chloros** is inseparable from its more general sense in which it is used typically, to describe fresh, green foliage. That is to say, the Greek equivalent of a phrase that would be translated into English as 'green foliage' is not ambiguous: when it is used descriptively, and with its prototypical meaning, in contexts such as this, the Greek **chloros** is not primarily a colour-term. (Incidentally, this was one of the words of which Gladstone said that they 'have very slight claims to be treated as adjectives of definite colour'.) In this respect it is similar to, though it may not match exactly, the Hanunoo word, **malatuy**; like the latter, it can be used to describe plants, fruits and other things that are yellow rather than green in colour. But when it is used referentially by philosophers who are discussing colour as such, it can be very plausibly said to have BK-green as its focal, or prototypical, denotation.

So, too, can the other two words that I have mentioned: **prasinos** and **poodes**. (In saying this I am aware that we might come to a different conclusion if we relied on the standard dictionary definition of these terms or took on trust what has been said by commentators, many of whom may have fallen victim to the etymological fallacy and the myth of literal meaning.) There are contexts, perhaps, in which a more specific shade of green is induced by contrast. But the three words I have mentioned all seem to be level-1 terms and, in contexts where colours are referred to and listed, equally general.

The fact that there is no single word for red or green (coupled with the fact that one of the words for green, **chloros**, does not normally denote colour independently of texture when it is used attributively, rather than referentially) may not go against the spirit of the BK-hypothesis, but it certainly invalidates some of the more simplistic and popular formulations of it.

I will not discuss in detail here other level-1 colour-terms in Ancient Greek, but I should perhaps mention that, although **xanthos** is readily identified as the word for BK-yellow (and would be on any list of BK-basic Ancient Greek colour-terms), it is not at all clear that Ancient Greek had a word for the sixth colour in the BK-hierarchy, blue. In fact, there are serious, and perhaps insoluble, problems relating to Ancient Greek words that denote colours in the blue-purple area of the spectrum. At least three, and possibly four, words have to be considered as basic level-1 colour-terms: **halourgos** (usually translated 'purple'), **kuanous** (which is of course the source of the English term **cyan** and is usually translated 'dark blue'), **ophninos** ('violet'?) and the famous **glaukos** (a word which is notoriously problematical when it is used attributively, is highly context-dependent and is variously translated as 'light blue', or 'grey-blue' or 'blue-green', in contexts in which it is used referentially of a colour which is clearly distinct from the others I have listed). Of these four, **halourgos** seems to have been the most salient. This is not what we would predict from the BK-hypothesis (on the assumption that the translation 'purple' correctly reflects the focal denotation of **halourgos**).

Another point at which the BK-hypothesis, as it is usually presented, appears to be invalidated by Ancient Greek is that, although Ancient Greek has, arguably, several words for blue, it has no equally basic word for BK-pink or BK-brown (unless **purros**, prototypically used of hair and variously translated as 'chestnut', 'orange', 'tawny' or 'reddish', referred focally to brown in contexts in which it is contrasted with other colour-terms). But, as I have already mentioned, there are great difficulties in specifying the denotation of several level-1 Greek words, even when they are being used contrastively and referentially as colour-terms.

Conclusions

I must now bring some of the threads together and recapitulate the major points that I have sought to establish in this chapter.

1. There is a crucial distinction to be drawn between the attributive and the referential use of colour-terms.
2. Reference to colour and colours is much less common in some cultures than it is in others, and in many it is probably a highly artificial practice.
3. The ability to refer to colour and colours, ostensively as well as non-ostensively, is facilitated – to put it no stronger – by having available for the purpose a language that has a word meaning 'colour' and other such abstract nouns, or to be more precise other such second-order extensional nouns, both count-nouns and mass-nouns.
4. Many languages of the world do not have a word meaning 'colour', and many languages do not have as well-developed a second-order (extensional or intensional) vocabulary as the languages that have been used over the centuries for philosophical and scientific discussion.

5. The richness of the vocabulary of colour-referring terms in many familiar languages is undoubtedly and demonstrably the product of culture, of which science and philosophy, no less than painting, pottery or weaving are a part.

6. Many languages of the world do not have colour-referring terms as such. They may, however, have words that are used to describe the appearance of things in terms of what we would identify as colour. These words are often context-dependent (as are many of the basic colour-terms in many familiar languages), but they may be used, if required, as several Greek words were, such as *chloros*, to refer to colours, and in such cases they may very well support the BK-hypothesis. But the definition of 'basic' by Berlin and Kay would have excluded them as basic colour-terms because of their context-dependence.

7. Finally, evidence from the acquisition of colour vocabulary by children, coupled with the fact that colour is not grammaticalised across the languages of the world as fully or as centrally as shape, size, space, time, etc., suggests that, for whatever reason, though perceptually salient, it is not linguistically salient unless and until it is made so by particular languages in consequence of cultural developments that over time have directly or indirectly created an enriched vocabulary of colour. Colours then, as we know them, are the product of language under the influence of culture.

Further reading

Berlin, B., and Kay, P., *Basic Color Terms: Their Universality and Evolution*, Berkeley: University of California Press, 1969.

Bornstein, M. H., 'On the development of colour naming in young children: data and theory', *Brain and Language,* **26** (1985), 72–93.

Burnell, D. (ed.), *Vesuvius and Other Latin Plays*, Cambridge: Cambridge University Press.

Clark, E. V., and Clark, H. H., *Psychology and Language*, New York: Harcourt Brace Jovanovich, 1977.

Conklin, H. C., 'Hanunoo color terms', *Southwestern Journal of Anthropology,* **11** (1955), 339–44 (reprinted in *Language in Culture and Society: Reader in Linguistics and Anthropology*, ed D. Hymes, New York: Evanston; London: Harper & Row, 1964.)

Corbett, G. G., and Morgan, G., 'Colour terms in Russian: reflections of typological constraints in a single language', *Journal of Linguistics,* **24** (1988), 31–64.

Crystal, D., *The Cambridge Encyclopedia of Language*, Cambridge: Cambridge University Press, 1987.

Davidoff, J., *Cognition through color*, Cambridge, MA: MIT Press, 1991.

Forbes, I., 'The terms *brun* and *marron* in modern standard French', *Journal of Linguistics*, **15** (1979), 295–305.

Gladstone, W. E., 'Homer's perceptions and use of colour'. In *Studies on Homer and the Homeric Age*, vol. 3, Oxford: Oxford University Press, 1858.

Grice, H. P., *Studies in the Way of Words*, Cambridge, MA: Harvard University Press, 1989.

Guthrie, W. K., *History of Greek Philosophy*, vol. 2, Cambridge: Cambridge University Press, 1965.

Kay, P., and McDaniel, C. R., 'The linguistic significance of the meaning of basic colour terms', *Language*, **54** (1978), 610–46.

Lucy, J. A., *Language Diversity and Thought: A Reformulation of the Linguistic Relativity Hypothesis*, Cambridge: Cambridge University Press, 1992.

Lyons, J., *Introduction to Theoretical Linguistics*, London, New York: Cambridge University Press, 1968.

Lyons, J., *Semantics*, 2 vols., Cambridge: Cambridge University Press, 1977.

Lyons, J., *Language, Meaning and Context*, London: Fontana/Collins, 1981.

MacLaury, R. E., 'From brightness to hue', *Current Anthropology*, **33** (1992), 137–86.

Morgan, G., and Corbett, G. G., 'Russian colour terms: problems and hypotheses', *Russian Linguistics*, **13** (1989), 125–41.

Moss, A. E., 'Basic colour term salience'. *Lingua*, **78** (1989), 319–20.

Platnauer, M., 'Greek colour-perception', *Classical Quarterly*, **15** (1921), 153–62.

Tornay, S. (ed.), *Voir et nommer les coulours*, Nanterre: Laboratoire d'Ethnologie et Sociologie, 1978.

Whorf, B. L., *Language, Thought and Reality*, Cambridge, MA: MIT Press, 1956.

Wyler, S., *Colour and Language: Colour Terms in English*, Tübingen: Gunter Narr, 1992.

Notes on Contributors

Denis Baylor is a visual physiologist. After receiving his medical training at Yale he has spent his career in research on visual mechanisms in the retina, where his particular interests are the rod and cone photo-receptors. He is a member of the National Academy of Sciences, and is Professor (and formerly Chairman) of the Department of Neurobiology at Stanford University.

David Bomford is Senior Restorer of Paintings at The National Gallery, London, where he has worked since 1968. He has lectured and published widely on the techniques of European painting and was co-organiser and co-author of the award-winning *Art in the Making* exhibitions and catalogues at the National Gallery. For ten years he was editor of the international journal *Studies in Conservation*. He is Slade Professor of Fine Art at Oxford University for 1996–97.

Janine Bourriau is a Fellow of the McDonald Institute for Archaeological Research, University of Cambridge and a Fellow of Darwin College. She worked in the Fitzwilliam Museum, Cambridge, where she organised two major exhibitions of Egyptian Art, and now concentrates on fieldwork and research. Her special interest is the art and archaeology of Egypt in the Second Millennium B.C., and her most recent book is *An Introduction to Ancient Egyptian Pottery*.

John Gage is Reader in the History of Western Art at Cambridge and a Fellow of the British Academy. His several studies of colour include *Colour in Turner: Poetry and Truth*, 1969, *Colour and Culture: Practice and Meaning from Antiquity to Abstraction*, 1993 and the forthcoming *Deconstructing Colour*.

Trevor Lamb, FRS, is Professor of Neuroscience in the Department of Physiology, University of Cambridge, and a Fellow of Darwin College. Originally an electrical engineer, he now studies the first events in vision. His main research interest has been the biochemical reactions which underlie the electrical response of rod and cone photo-receptor cells to illumination, and he has also investigated the communication of visual information between cells within the retina.

Malcolm Longair is the Jacksonian Professor of Natural Philosophy at the Cavendish Laboratory, University of Cambridge. From 1980 to 1990, he held the joint positions of Astronomer Royal for Scotland, Regius Professor of Astronomy at the University of Edinburgh and Director of the Royal Observatory, Edinburgh. In 1990, he delivered the Royal Institution Christmas lectures on television on the subject of 'The Origins of our Universe'. His research interests are centred upon high energy astrophysics and astrophysical cosmology. His recent publications included *Our Evolving Universe* (1996). His scientific interests include the history of physics and astrophysics and he published a history of twentieth-century astrophysics and cosmology in 1996.

John Lyons, Master of Trinity Hall, Cambridge, since 1984, has previously been Professor of Linguistics at the Universities of Edinburgh and Sussex. He is a Fellow of the British Academy, has been awarded honorary degrees from several British and foreign universities, and in 1987 was knighted for services to linguistics. His major works include *Introduction to Theoretical Linguistics* (1968) and *Semantics* (1977).

John Mollon is Reader in Experimental Psychology in the University of Cambridge. His main research interests are in visual perception and he has made a special study of colour, in its phenomenological, psychophysical, physiological, and molecular genetic aspects. He is a former Chairman of the Colour Group of Great Britain and a former Honorary Secretary of the Experimental Psychology Society.

Peter Parks is a director and producer of films, and specializes in the diverse fields of marine biology and special effects. He co-founded Oxford Scientific Films and subsequently formed his own production company, Image Quest. A full member of the British Society of Cinematographers, he is currently producing a trilogy of large format (70 mm IMAX) aquatic movies filmed on the Great Barrier Reef.

Bridget Riley is a practising abstract painter. She has been exhibiting internationally since the early 1960s, when she first won recognition for her black-and-white paintings. She was the first British painter to win the International Prize for Painting at the Venice Biennale. With Peter Sedgley she set up SPACE, a plan for artists' studios in London. Bridget Riley was a trustee of the National Gallery during the 1980s and in 1992 she showed her most recent colour work at the Hayward Gallery in an exhibition entitled 'According to Sensation: Paintings of the Last Ten Years'.

Acknowledgements

CHAPTER 1

Figures 1, 2, 3, 5, 6 and 8 The National Gallery, London.

Figure 4 Galleria degli Uffizi, Florence.

Figure 7 Musée d'Orsay, Paris.

Figure 9 Sterling and Francine Clark Art Institute, Williamstown, Massachusetts.

CHAPTER 2

Figures 1 and 4 Museo del Prado, Madrid.

Figure 2 Vatican Museums.

Figures 3 and 7 The National Gallery, London.

Figure 5 Rijksmuseum-Stichting, Amsterdam.

Figure 6 Hermitage Museum, St Petersburg/The Bridgeman Art Library.

Figure 8 The Art Museum, Princeton. From the collection of William Church Osborn, Class of 1883, Trustee of Princeton University (1914–1951), President of the Metropolitan Museum of Art (1941–1947); gift of his family.

Figure 9 and 10 The Museum of Modern Art, New York.

CHAPTER 3

Figures 4(a) from Fauval, J., Flood, R., Shortland, M., and Wilson, R. (eds.), *Let Newton be: A New Perspective on his Life and Works*, p. 87, Oxford: Oxford University Press, 1989. Reprinted by permission of Oxford University Press..

Figure 6(a), 7 and 8 from Greenler, R., *Rainbows, Halos and Glories*, Cambridge: Cambridge University Press, 1991.

Figure 11(a) Courtesy of the Cavendish Laboratory.

Figure 13 from Williamson, S. J., and Cummins, H. Z., *Light and Colour in Nature and Art*, Plate 10 between pages 201 and 202, New York: John Wiley and Sons, 1983. Copyright ©1983 reprinted by permission of John Wiley and Sons Inc.

Figure 14 from Maxwell, J. C., 'On physical lines of force. II. The theory of molecular vortices applied to electric currents', *Philosophical Magazine*, **21** (1861), 4th series, Plate V, Figure 2.

Figure 15 from Hertz, H., *Electric Waves*, London: Macmillan and Co., 1893.

Figure 16 from Millikan, R. A., 'A direct photoelectric determination of Planck's "*h*"', *Physical Review*, **7** (1916), p. 373, Figure 6

All other figures are from originals by M. Longair.

CHAPTER 4

Figure 2 from Dowling, J. E., and Boycott, B. B., 'Organization of the primate retina: electron microscopy', *Proceedings of the Royal Society of London*, **166**B (1966), 80–111.

Figure 3 after a drawing by H. Saibil.

Figure 6 after Kraft, J. W., Makino, C. L., Mathies, R. A., Lugtenburg, J., Schnapf, J. L., and Baylor, D. A., 'Cone excitations and color vision', *Cold Spring Harbor Symposia on Quantitative Biology*, **55** (1990), 635–41.

Figure 7 after Baylor, D. A., Nunn, B. J., and Schnapf, J. L., 'Spectral sensitivity of cones of the monkey *Macaca fasicularis*', *Journal of Physiology*, **390** (1987), 145–60.

Figure 8 from Nathans, J., 'The genes for colour vision', *Scientific American*, **260**, 42–9. © Scientific American Inc., George V. Kelvin.

Figure 9 reprinted with permission from Nathans, J., Thomas, D., and Hogness, D. S., 'Molecular genetics of human color vision: the genes encoding blue, green and red pigments', *Science*, **232** (1986), 193–202. Copyright 1986 American Association for the Advancement of Science.

Figure 11 from DeVries, S. H., and Baylor, D. A., 'Synaptic circuitry of the retina and olfactory bulb', *Cell*, **72**/*Neuron*, **10** (suppl.) (1993), 139–49. Copyright Cell Press.

All other figures are from originals by D. A. Baylor.

CHAPTER 5

Figure 1 from Ferdinand Hamburger, Jr, Archives of the Johns Hopkins University.

All other figures are from originals by J. Mollon.

CHAPTER 6

All figures are courtesy of P. Parks.

CHAPTER 7

Figure 1 from Albers, J., *Interaction of Color*, New Haven, CN: Yale University Press, 1963 (paperback abridgement, 1971). Courtesy of the Josef and Anni Albers Foundation.

Figure 2 from Field, G., *Chromatography, or a Treatise on Colours and Pigments and their Powers in Painting*, 2nd edn, 1841.

Figure 3 from Maerz, A., and Paul, M. R., *A Dictionary of Colour*, 3rd edn, New York: McGraw-Hill, 1953.

Figure 4 from Schiffermüller, I., *Versuch eines Farbensystems*, coloured fold-out opposite p. 14, Vienna, 1772.

Figure 5 from Syme, P., *A Nomenclature of Colours*, 1821.

Figure 6 from Nicholson, W., 'Liberation of colour'., *World Review*, 1944 (reprinted in *Unknown Colour*, London: Faber, 1987).

Index